SOILS AND FERTILIZERS

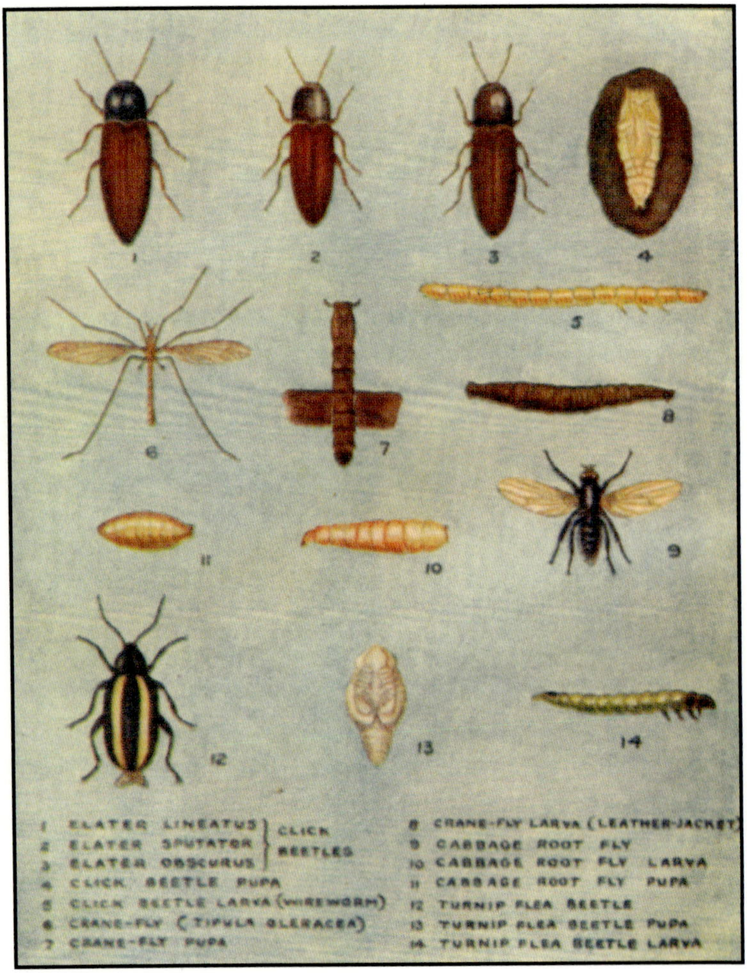

WIREWORM AND OTHER GROUND PESTS

SOILS AND FERTILIZERS

BY

A.J. MACSELF

ILLUSTRATED IN COLOUR
HALF-TONE AND LINES

SATISH SERIAL PUBLISHING HOUSE

SSPH

403, Express Tower, Commercial Complex, Azadpur, Delhi - 110033 (INDIA)
Phone : 011-27672469, 27672852 Fax : 91-11-27672046
E-mail : ssp@ndf.vsnl.net.in, hkjain@satishserial.com
Website : www.satishserial.com

Published by :

SATISH SERIAL PUBLISHING HOUSE
403, Express Tower, Commercial Complex, Azadpur, Delhi - 110033 (INDIA)
Phone : 011-27672469, 27672852 Fax : 91-11-27672046
E-mail : ssp@ndf.vsnl.net.in, hkjain@satishserial.com

First Indian Reprint 2005

Printed at
SALASAR IMAGING SYSTEMS, DELHI-35

PREFACE

IN the several books on horticultural subjects that I have written some reference has of necessity been made to soils, their varying character, the methods by which they should be cultivated or improved, and to the important functions of manures and the many diverse substances which are embraced by the general term "Fertilizers."

Whilst such references have always been recognized as indispensable there has invariably been a feeling that their enforced brevity has rendered the paragraphs or the chapter seriously inadequate, and perhaps calculated to convey an erroneous impression regarding the real importance of soils and their management, for in very truth it is hopelessly impossible to attain success in the making or maintenance of a garden unless this theme is made the subject of close and constant study.

Scientists, chemists, and writers of undoubted talent have written much upon soils and upon manures or fertilizers. Their writings are learned and profound; the subject affords grand scope for the chemist, and the tutor, but between the student of the class-room or the laboratory and the plain working gardener or the amateur whose garden can occupy but a fraction of time and thought exists a wide gulf, and that which is plain teaching to the former may be and frequently is quite above the heads of the latter. It is the latter class, however, whom I strive to serve and for whom I write in the hope they may find something of material aid; something which shall help them on the road to success in horticulture whether that be their chosen calling or their recreation and hobby.

My aim is therefore to write on a subject which is essentially scientific, but to write in a style that he who runs may read, and conscious as I am of the difficulty of my task, it is faced and tackled in consequence of the conviction that a plain, simple book on the subject is a real need.

Whether my work successfully meets the need cannot presume to judge.

Of my critics I ask no mercy beyond that they may acknowledge honesty of purpose in my effort.

THE AUTHOR.

CONTENTS

LIST OF ILLUSTRATIONS

COLOUR

HALF-TONE

LINE

SOILS AND FERTILIZERS

CHAPTER I

MOTHER EARTH

DEPENDENT as we are throughout our mortal life upon the earth and its products, we may surely consider time well spent which is occupied in studying the earth from whence we sprung, upon which we live, and by which we are sustained. Strange indeed must be the mentality of the person who would dismiss the subject as being dry and uninteresting. Fool he would be who claimed that he had nothing to learn from the earth, and the wiser the student the more readily he admits how little is his knowledge.

The earth is to every man the world upon which he lives and from which he manages by divers means to obtain sustenance and comfort; it is the farmer and the gardener who speak of the soil. What is the soil ?

Scientists speak of silica, alumina, potash, lime, magnesia and various other elements or constituents of soils; the non-scientific husbandman thinks of the soil as the earth's crust; that superficial layer of mcvable and workable earth, varying considerably in thickness and in texture or character which he can cultivate and make productive of the necessities of life.

The farmer and the gardener give their own simple names to soils, and speak of light soil, heavy soil, rich soil, hungry soil, clay soil, or sandy soil. For each of these and similar terms there is a scientific appellation, and were it only possible for all who cultivate the soil to take their degree in agricultural science, chemistry and botany, the exclusive use of scientific names and terms would prove to be an immense advantage.

Crying for the moon is no more futile than it is to bewail the fact that all men are not scientists, and we may for the present be better occupied in seeking a clear understanding regarding the accepted meaning of such terms as light soil, clay, loam, rich soil, or hungry soil.

Let there be no mistaken notion that because all cannot fully comprehend and make good use of scientific terms science is immaterial and valueless to practical horticulture. Even the rustic who will contemptuously declare that he "dont hold with that science at all," will proceed but a little way with a season's work on his vegetable plot without unconsciously adopting scientific principles with success that is partial only by reason of the ignorance which, by blinding him to the purpose and the reason, enslaves him to faulty and inaccurate methods.

In all probability the individual who is avowedly

non-scientific and proud of it will prove to be a more tractable pupil than one who has a mere smattering of science and is proud of it. A little knowledge is truly a dangerous thing when that little happens to be of scientific character, existing as a detached fragment in a mind that has had no grounding in science.

That is why the handbook written in scientific jargon for the untrained reader is very prone to work mischief, and it is also why the present volume tells the story of soils, their management and improvement in the plainest and simplest language.

Reverting to the question, "What is soil?" The simple reply that it is the earth's crust is not adequate answer, as the cultivator will realise when he delves at various times in soils of widely differing character.

The strange thing is that, widely as soils differ in appearance and texture, they are practically all the product of combinations of the same group of substances or ingredients, but it is variation of balance or proportion in which these ingredients are mixed, and sometimes the absence of certain of them which produces the striking differences In appearance.

At some remote period of the earth's existence it is reasonably to be supposed that its surface was bare, hard rock with no covering crust of soil.

We are fairly acquainted with the fact that continuous atmospheric changes, driving soaking rains, drying winds, baking sunshine, and keen frosts have an effect upon the surface of even hard, forbidding rocks. Gradually, particle by particle, the face of a rock is disintegrated, and crumbles away in the form of small stones and grit. At times heavy rains, at others rapidly melting snow and ice will cause torrent or avalanche, and in the helter-skelter rush of water down the mountain side the loosened fragments of rock are carried down until they reach a valley or depression in which they settle, and in course of time the valley or the hollow becomes levelled up with a mass of pulverised rock, and thus is created soil. At the present time the mountain torrents and avalanches carry with them much; in addition to stones and grit; leaves, stems, roots of plants, carcases of animals, reptiles, insects; birds, their feathers, their excrement, and their eggs and nests are hurtled to destruction, mingled with the broken rock and pulverised grit, and being buried, they decompose, adding different substances to the mass which eventually becomes a layer or stratum of a different kind of soil.

By making deep excavations varying strata are frequently disclosed which enable the era of their formation to be determined. Naturally, when rocks were hard and barren, destitute of any form of vegetation the deposits of disintegrated rock formed were entirely of mineral or stony character. At later periods vegetable and animal matter mingled with the stony debris and produced softer soil.

Pulverised stone alone, free from vegetable and animal matter, became sand, for sand is stone in minute particles.

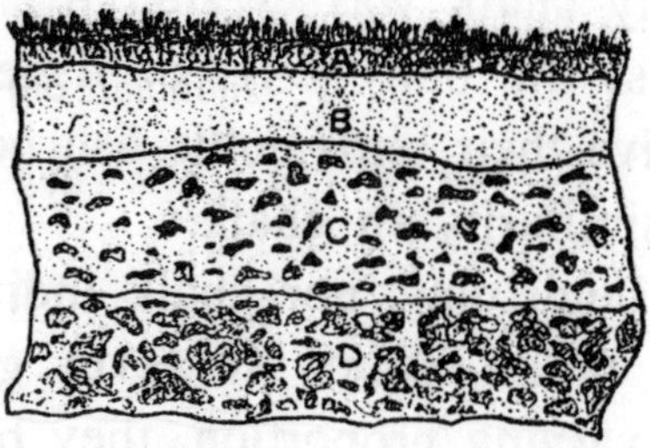

SOIL STRATA

A, Turf. B, Top spit of fertile soil. C, Second spit, which must be enriched. D, Rocky subsoil which should be well broken up with a fork.

Clay is also formed by the pulverization of rock, but the rocks which produce clay are of a different character from those which produce sand. Slate, granite, and other stones which will grind down to very fine powder form clay, the minute particles in association with water becoming so tightly compressed that instead of forming sand of finer grain or texture they form solid plastic masses through which neither air, nor water can percolate. Generally, however, a certain amount of vegetable or animal matter is found as well as silica (stone) and water, and although such a mixture is termed "impure" clay it is of far greater value from the standpoint of the cultivator than pure clay.

Chalk is frequently found in dense, solid masses, mole or less freely interspersed with flints and fossils of various kinds. The scientific name for chalk is calcium carbonate, and it is otherwise known as carbonate of lime.

Originally chalk was created from sea-shells, and doubtless the bony parts of fishes. Bird fanciers who give their birds cuttle-fish bone will recognize its similarity to chalk.

Neither sand, clay, nor chalk individually and alone will make what we term garden soil. Blended in varying proportion, they become more or less capable of serving as a rooting medium for some kind of vegetation, but a mixture of these three, assuming each to be pure, would not make a fertile soil.

Humus is required as an additional ingredient, and its importance is so great, that we must pause to consider its character.

Humus is really rotted matter formed of leaves, stems, or roots of plants, including the bark or wood of trees and shrubs, the stubble or straw of corn, hay, or green grass. Decomposed flesh, bones, skin or fur of animals and fishes or of birds, including their feathers, are capable of forming humus, as also to a varying extent will dung, wool, hair, and most things that will rot in the ground.

Humus enriches the on during the process of decomposition, and apart from actual plant food it supplies during that period, the residue is of an absorbent nature, capable of holding moisture, and incidentally liquid nourishment, as it dissolves in the soil.

If one fills a plant pot with sand and then waters it, the bulk of the water will pass through the drainage hole of the pot and the sand will rapidly regain a dry state. If, however, a dry sponge is cut in small pieces and these are evenly distributed among the sand before water is given the pieces of sponge will soak up and hold a great deal of the moisture, and as the sand tends to become dry the sponge will gradually release its moisture, thus considerably prolonging the period during which the sand will remain moist. In much the same way humus will act upon a light, sandy soil, greatly to the benefit of the plants growing therein.

Furthermore, humus acts as a warming agent to soils that are otherwise cold. Sand, clay and chalk soils that lack humus are always of a light color. Humus is usually dark brown or nearly black, and the addition of humus to a light coloured soil will darken it. The darker a soil is the more heat it will absorb from the sun and the longer it will retain the heat absorbed. It is not, of course, due to the changing of the colour alone that the power of retention of heat is affected, but it may suffice to show that whilst soil is darkened by the addition of humus, so also is its power of absorption and retention of heat increased.

Humus has been the subject of much pen-work. Scientists have written a great deal about it, and much that one has written has been queried or flatly contradicted by another.

For instance, some writers have affirmed that plants are capable of actually absorbing humus or humic acid from water in which humus has been steeped. Others have denied that the truth of these statements has been proved, and these maintain that humus merely absorbs and holds available for plants the dissolved plant foods derivable from other substances. In a book which was strongly recommended to me when I was first seeking knowledge of soil management about ten pages of matter dealt with humus, its chemical nature, its absorbent power and its value, and then in a paragraph of half a dozen lines the mystifying statement was made that "Humus is not indispensable to the life or full development of plants." The only reason given for that statement was that experiments had proved that carbonic acid gas is of itself able to supply plants with the carbon necessary to promote growth. That is doubtless an interesting fact to the scientist of the laboratory, but it is not intended as a piece of advice to the farmer that instead of bothering himself about the supply of humus in the soil of his wheat fields and potato fields he should simply contrive by some fantastic means to discharge into them quantities of carbonic acid gas.

Unfortunately, some who rely upon reading a few scientific works to enable them to assume the role of tutors get hold of fragmentary statements and endeavour to build up what they term practical instructions from them, disseminating these instructions all unconscious of the mischief they may work. Thus it has happened that writers have scattered abroad statements which conveyed the impression that the more humus a soil could be made to contain the better plants and crops would grow. Such is an utterly erroneous idea, for quite a large proportion of garden soils contain vastly more humus than is good for the great majority of plants. So also in regard to methods advised for augmenting the supply of humus in soil a deal of harm is frequently done.

The simple statement that humus is derived from decayed vegetable matter has prompted many to urge that any and every kind of garden refuse, leaves, and rubbish should be buried in the trenches as digging proceeds. This is most dangerous advice, antagonistic to hygiene of the soil, and has been responsible for an immense amount of trouble with a host of plant pests and diseases.

One sees all kinds of diseased, pest-infested vegetables, fruits, plants and garbage thrown first on the rubbish heap to rot, and then wheeled out on the garden to be dug in, spreading spores or mycelium, germs or eggs of diseases and plagues from one plot to all the plots or beds in the garden, all with the mistaken notion that benefit alone can result from the addition of humus to the soil. The time is long overdue when such a practice should be abandoned, for it is unscientific, unclean, and destructive of healthy soil, and comparable only to the ignorance and folly which would deliberately ignore the dangers of filth and pestilence in other departments of the home. An incinerator should be the destination of garden refuse, the ashes from which constitute clean wholesome and rich fertilizer.

When humus is lacking, it may readily be supplied by vastly better methods which will presently be described. The prior consideration is what proportion of humus is good, and what soils require to be supplied with it.

The customary plan when writing upon the characters of soils is to describe methods of conducting what is termed mechanical analysis of soils, and sometimes chemical analysis is also advised and the process described. It is not my intention to follow this course. In the first place few readers would be likely to follow advice to analyses soil; secondly, such analysis is more likely to be grievously inaccurate and misleading than helpful; and thirdly, with a little very ordinary intelligence the mechanical character of a soil may be nearly enough determined by examination and handling to indicate the direction in which improvement is desirable.

It is seldom necessary, for instance, to sift a sample of soil through various sieves and weigh the different grades of siftings in order to ascertain whether it contains a large or small proportion of stone. Even the budding amateur gardener can tell by handling whether a spadeful of soil is a solid mass of tenacious clay or a loose porous sand. Blackish or dark brown colour will suffice to show that humus is present and that a whitish or pale yellowish brown soil cannot be liberally supplied with humus, and although it is necessary that one shall ascertain the balance of proportion in a soil of mixed character it is by no means a certainty that an amateurish effort to analyse it with such apparatus as the average garden owner may have at hand will result in anything like accuracy.

Chemical analysis is an altogether different proposition from mechanical analysis. It requires not only proper equipment but an extensive knowledge of chemistry.

Even with both it is an open question whether analysis is in the great majority of cases of real practical value. The general custom is to wander over a plot of ground taking an odd shovelful of soil here and there at random, mix together, weigh off a given quantity and proceed to submit this sample to minute analysis. How anyone with a knowledge of science can suppose that the result of such a procedure can give an accurate or even approximate idea of the actual contents of the soil of an acre, a rood, or even a pole of ground is beyond comprehension.

It would be necessary to repeat the whole process twenty, fifty, perhaps a hundred times, and compare results before the truth would be arrived at, and if the ground is dug once, twice, thrice, the analysis would show wider variation and prove no more than that digging had improved the soil.

Analysis conducted by fully qualified analytical chemists with all the necessary apparatus and facilities, and with samples taken and prepared under their personal supervision, may be and has been the means of increasing knowledge and of rendering useful assistance in the general advance of soil science, but the average garden owner and garden worker will be more profitably occupied by first examining his soil, judging by its appearance and handling whether it is clay, deficient in both sand and humus; sand, deficient in humus and clay, or a soil that is rank with humus and requiring what are commonly called correctives or soil sweeteners.

The ideal garden soil is a well balanced mixture of clay, sand, chalk, humus, and stone. Other ingrediënts there will be, some essential and beneficial, some immaterial, and probably some detrimental, but at the moment these five are sufficient to engage our attention. It must not be supposed that equal proportions of the five constituents would be by any means a well balanced soil. It is considered a good balance if about 50 per cent. of the bulk is sand, 30 per cent. clay, with not less than 7 nor more than 10 per cent. humus, and the rest is made up of moderately small stones and chalk or lime. The perfection of even such a balance depends to a certain extent upon auxiliary conditions; elevation and exposure, character of subsoil, drainage, and average rainfall of the district render variations in these proportions desirable, and sometimes considerably important, and when, after generalising, we come to consider ideal soils for particular plants or crops the necessity for modifications becomes increasingly urgent. Seldom will a soil merit description as ideal garden soil when first it is taken under cultivation. Many skillfully cultivated gardens may reasonably be described as ideal, and yet, if several of these were examined or even submitted to most accurate mechanical analysis, they would reveal startling variations in character or texture. To attempt, therefore, to tabulate a formula for an ideal garden soil and represent it as appropriate and accurate for all localities, situations and purposes of gardening is neither scientific nor practicable.

Mother earth is very wonderful and very responsive; generous beyond compare to those who lend their time, their interest, their labour; safer than banks to yield liberal interest upon judicious investments; greatest of all producers both of the necessities and joys of life.

CHAPTER II

FIRST STEPS IN SOIL IMPROVEMENT

ASSUMING that a plot of land which has never within knowledge been cultivated as a garden is to be taken in hand, it is important to realise at the outset that haste is dangerous, and may easily prove disastrous. All too often one se s a newly built house occupied in spring, a garden designed, made and planted within a few weeks, and usually spoilt for years to come.

Impatience is a greater enemy of good gardening than greenfly or wireworm, and is far more difficult to fight. It is seldom the most prudent plan to begin by ordering a liberal supply of manure and proceed to dig it in. This may happen to be right, but investigation is necessary to ascertain the fact, and very rarely will one be wrong in deciding that manuring may be delayed until the second, and perhaps even the third digging.

Should the plot be covered with turf, the first step should be to remove the turf, cutting the sod fairly thin and stacking it in a convenient spot. Sometimes it is recommended that turf should be buried, but usually it proves to be a nuisance, not only because many hard rooted weeds, such as dandelion, dock, couch grass, plantain and butter cup will be distributed throughout the plot, but because large chunks of turf, especially if very dense and full of fibre, will occupy a long time in decomposing. True, the fibre and soil of a clean healthy turf is the most valuable part of a piece of meadow land, and it should not be wasted; but in view of the above-mentioned risk of trouble and of the further likelihood of injurious ground pests being present, it is safer to stack the turf, dressing each layer as building proceeds with a soil fumigant; then when well decomposed the loam may be evenly distributed and dug into the top spit where it will do most good. If the plot is merely reclaimed waste or an old neglected garden, the first step should be to clear the surface of all rubbish. Builders are prone to leave all manner of debris on the ground around the building they have erected, and all such material as broken slates, tiles, wood, and particularly glads, should be removed.

When the surface is cleared take note of its contour; if very uneven, considerable difference may be expected between the soils of the hollows and the higher levels, and to ascertain this holes should be dug in various places, and if possible to a depth of at least three feet. Watch very carefully during the process of opening each hole for variations in the substance, solidity and colour of the soil, for herein will be found a most important guide to further procedure. One may find a top crust of brown earth resting upon a closer lighter coloured sandy soil. If the top crust is a foot or more in depth and the lighter sand extends to the full three feet, one may congratulate himself upon having a soil amenable to simple treatment.

Perhaps the top spit will be a yellowish-brown clay which cuts almost like cheese or soap when moist or is hard as bricks when dry, the change as one descends being toward a tougher, sticky texture, a lighter yellow, with sometimes streaks of bluish or leaden tint, and scarcely any stone. Such is a soil that will require a great deal of working, but which will pay handsomely for every bit of labour put into it.

One may have to deal with a red or yellow gravel, with a considerable number of stones near the surface, and more stone the deeper one goes, or the gravel may suddenly give way to solid, white chalk with more or less of flint stone embedded among the chalk. Such soils are not to be coveted by any garden lover, for they are ever ravenous for nourishment, and even with liberal treatment will only attain suitability for certain classes of plants. The clever gardener with such a soil will recognise his limitations and will decide to confine his attention to naturally suitable or adaptable classes of plants; by so doing a very enjoyable garden may be formed even on chalk downs and in old chalk pits. Heathland or moorland peat, intermixed as a rule with fine powdery, dry and dusty sand is another possibility, and an unwelcome one where general gardening is the object in view, but here again there are classes of plants and styles of gardening for which such soils are peculiarly adapted.

The loose, hot sand of the seaside, frequently lying on hard rock, the gritty, porous mixture of rotted stone and debris to be found in districts where sandstone hills abound, and the soft black sands and mossy peats of marshes and estuaries of rivers and of the fens are localised possibilities, all of which must claim attention and receive special treatment.

When the three-feet holes are opened three separate points must be observed, for each is of distinct and great importance. First, observe the difference in the appearance of the soil taken from the first, second and third foot depth of a hole, for the subsoil as well as top-spit is of great concern to the horticulturist. Secondly, compare the soil from holes in different parts of the plot. It may be found that an interchange is calculated to be beneficial, at, on the other hand, that one part at the garden is specially adapted for one class of plants and another part for something totally different, and the design of the garden and treatment of its varied soils will be arranged accordingly.

The third point to note is whether and how rapidly water appears in the holes, whether in all holes to about an equal depth or whether some holes become flooded and others not. This will afford some clue to the condition of drainage, and that is one of the most important factors in soil management which in due Course will be afforded careful consideration. From information gathered up to this point it may be determined whether the subsoil is lacking in porosity, or on the contrary, is too largely composed of stone and sand, and would be improved by incorporation of clay or strong loam. In all probability the addition of humus - forming substance of some kind will be desirable, for a subsoil which has long remained undisturbed is unlikely to be well supplied with humus. Without the slightest hesitation the owner of a clay subsoil may procure leaves, grass, chaff, spent hops, grains, for the purpose of providing humus for the subsoil, also sharp sand, mortar or lime rubble, river or sea-sand, old coal or coke ashes for improving porosity. There is sometimes an advantage in burning some of the clay itself, distributing the pulverised burnt earth through the subsoil as trenching proceeds. Some of the turf which is pared from the surface may be burnt and will provide substantial benefit upon a cold and tenacious subsoil.

Under present-day circumstances it is scarcely prudent to suggest that good stable or farmyard manure should be sacrificed to the first trenching of subsoil; this richer form of nourishment will with greater wisdom be reserved for the top spit.

The claying of a sandy or gravelly subsoil is not so simple a problem; there are circumstances where it may be done with considerable benefit, but in other cases there is les certain. Where the subsoil is so dry and Porous that a brief spell of drought tells heavily upon plants and crops an immense benefit follows a liberal incorporation of clay and peat or peat moss litter, but the mixture should undergo thorough preparation by exposure to frost, by liming, and frequent turning before being buried, and such labour is only practicable for comparatively small beds undergoing preparation for some particular purpose., A subsoil may be sandy or stony, and yet lie low and damp, and in such cases the addition of clay would tend to cause or aggravate stagnation.

Let the chief concern regarding agents of improvement in the first place be directed to the subsoil; the top spit will be accessible and can be nourished and improved at a second digging. One word of warning may here be necessary; if the surface is badly infested with noxious weeds of a perennial character the ground must be forked over and cleaned as well as possible before any effort is made to trench it.

Begin trenching as early in the autumn as possible. Trenching in middle summer cannot be recommended, for a man cannot reasonably be expected to dig in the heat of summer an area that would be quite comfortably managed in autumn or winter, and if the ground is dry and hard indifferent work and broken tools are inevitable. We will not linger here to describe the different methods of digging, trenching, and bastard trenching; they have their own chapter, as befits their importance, but emphasis must be laid upon the necessity for even and careful distribution of whatever is to be incorporated even with the sub-Soil.

Few errors are greater than that of supposing that the mere act of burying a good thing in the soil is satisfactory and beneficial. Large lumps, masses, layers of manure, clay, peat, humus, or even sand and grit should never be tolerated. Spread evenly, thinly, and mix as thoroughly as possible with the staple soil. That is not only sound advice, but must be a law inviolable to every cultivator.

After trenching and dressing the subsoil it will be advisable to allow a matter of two or three weeks for settling down, and then the top spit may be re-dug, this time with the purpose of adding and incorporating whatever it has been decided will improve it. One reservation must here be made, for if it is desired to add clay or marl to a sandy or gravel soil, the proper course will be to spread the dressing evenly over the surface and leave exposed to weather until the greater part of winter is over. It will be found that what was originally tough, sticky, plastic clay has been pulverised by the action of frosts, and crumbles down in fine particles. It is very important that a time shall be chosen for digging this in when the surface is wind dried for if the clay is wet, treading will undo the benefits of the frost's pulverisation by kneading the particles together again.

To dig during frost is not desirable when it involves burying hard frozen soil, but to dig in snow is worse, and no soil should be dug when wet enough to cling to one's boots and churn up the surface in pasty mud. Claying the top spit of a dry sandy soil or a harsh gravel is an excellent procedure, far more generally beneficial than claying the subsoil. It is desirable, however, that the clay shall be from a top spit, and if possible that it shall have been previously cultivated. Clay from a pit is inert, usually acid, and altogether unsuitable for garden purposes.

Manure also should be dug in the top spit, and if the soil is of light, poor quality cow manure or mixed farmyard manure will best meet the requirements, horse manure being more suitable for cold, heavy clays. Burnt earth, sand, old pounded lime rubble, leaf mould if available, spent hops, lawn mowings or for very stubborn clay a thin coating of sifted old and weathered coal ashes will all help to open up and improve the porosity of the soil, and it will also be beneficial to leave the surface in fairly rough clods instead of attempting to break them down. The greater the surface exposed while frosts are possible the more friable the soil will become by spring. Let that statement be emphasized, for often as it has been made one marvels at the carelessness of many cultivators who will leave land hard trodden and undug all winter and sacrifice the incomparable benefits of aeration and frost. Excessively hard clay soils may with great advantage be thrown up in steep ridges, thus exposing a much greater surface than is obtained by level digging. In spring the ridges will break down comfortably during spells of fine weather.

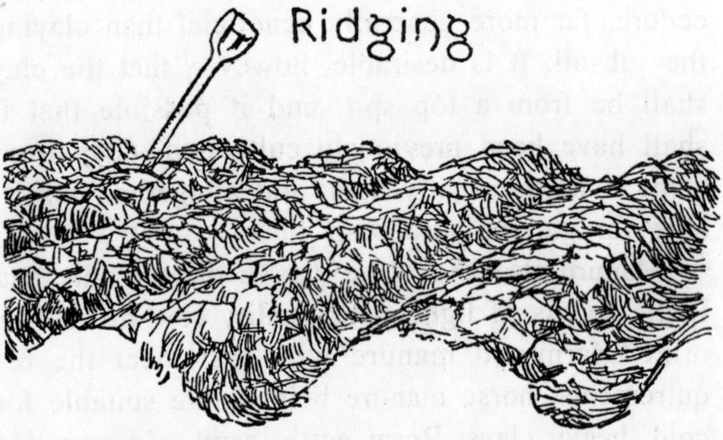

RIDGING

The ridges are made singly, working backwards, and throwing soil toward the centre of each ridge.

Whilst the beneficent effect of frost is made so obvious by this breaking down of stubborn clay it should not be imagined that lighter, and even sandy soils, have no need of similar exposure to frost. Light soils which have long been compressed and undisturbed become sour, and have need of the sweetening influence of both frost and wind, and rough digging also helps to free the soil of insect pests, many being killed by the frost and others devoured by birds. In this connection it is well to take due note of the fact that a piece of land dug, say, in mid-autumn, will settle down and be beaten to a close-knit surface by heavy rains quite early in winter. On this account prudent gardeners frequently express preference for mid-winter digging rather than mid-autumn, and presuming that a plot can be dug only once, and that mid-winter is kind enough to afford spells of open weather when the soil is not too wet and sticky for digging that preference is justified, but when opportunity allows it is better still to dig first in autumn and again in winter, and this is doubly advantageous when the soil has not previously been well cultivated.

MOORLAND PEAT LANDS and low-lying wet peats are more difficult to deal with than either of the foregoing soils if the aim is to convert them into gardens of general character.

For the purpose of specialising in rhododendrons, azaleas, heaths, ferns and peat-loving plants and shrubs of all kinds, fortune favours the possessor of a stretch of natural heath or moorland. His soil is already made, but when one is situated where all the soil is peat and he essays to make a rosegarden, herbaceous border, fruit plantation and kitchen garden he faces no simple task. Peat has little substance or strength, is deficient in most of the essential constituents of a good garden soil, is overcharged with humus which is of greater disadvantage because of the absence of lime. Usually peat lands are far removed from strong loams or useful clays, and even if such are accessible it is no simple matter to mix them effectively, because peat does not readily break down like other soils on account of the great amount of tough fibre of which it is largely composed.

Even frost has not the same action upon peat it has upon clay, for although it will loosen the solids, which will be very largely composed of fine sand, it does little toward disintegrating the tangled masses of fibre.

A good plan is to throw up the peat in rough chunks in autumn, then after frosts beat and shake the lumps to free the fibre from the finer matter; pitch the loose fibre into heaps and light a series of fire. The charred remains of the fibre will afterwards break down and should be evenly distributed and dug in with the sand and earthy matter. This will be one step toward creation of a good garden soil. Furthermore, the burning of the fibre will facilitate the more perfect blending of the staple soil with any imported loam or clay, and it will be really essential for some solid soil to be introduced before the garden will become fertile. Indeed, even though peat lands may be poor and hungry, it will not be so beneficial at the outset to apply animal manures as to apply good loam or clay. The manures will tend to increase the humus, already present in excessive quantity, whilst the clay or strong loam will help to solidify and tighten the soil. At best the conversion of such soils must be a work of time, and it is well to be content for a year or two to grow rank growing annuals and vegetables before investing too freely in more permanent subjects such as roses and fruit trees. Peat lands are practically certain, to be deficient in lime, and a dressing of ground lime will effect considerable improvement, or if the first breaking up is tackled in early autumn gas lime may be spread, and allowed to lie on the surface until spring. A hundredweight to four rod is the full quantity of gas lime that should be applied.

Growers of rhododendrons, heaths, and other shrubs which dislike lime will of course omit this treatment, and as a matter of fact their preparatory work will be much simplified, the chief thing being to break up the whole area to a good depth, and to work in manure, that from sheep, deer, or goats being most suitable for the purpose, by reason of the fact that they contain a greater proportion of soluble plant food than of humus, which the peat does not require. It is well that this should be done from four to six months before planting, and where fibre is found in excessive quantity a little burning will also be of advantage. If bracken is abundant as many of the rhizomes as possible should be dragged to the surface and burnt, thereafter keeping a close watch for the appearance of young fronds, which must, be cut off before they begin to unfold. A word of caution may be advisable to those who desire to make a lawn on peat lands. Efforts to improve the soil with a view to turfing with the usual meadow grasses are seldom successful; years of labour and much expense will be involved in establishing a lawn of grasses which belong to loams and what are termed heavy soils. It is infinitely better to be content to secure a good surface by weathering, burning, etc., but to leave liming, clay dressing and such efforts at changing the character of the soil severely alone, and then obtain a specially blended mixture of grasses which are suitable to peat lands, and happily there are such grasses, some of which would be quite unwelcome in gardens of average character, but with the customary regular mowing which lawns should have will succeed beautifully on peat.

The black sands or silts of fen districts are generally highly charged with humus, and although frequently flooded are so open and porous that they very rapidly drain. Attempts to improve these soils by introduction of marl and clay are not always successful, for if the situation is hollow and low lying the pasting of the soil is calculated to interfere with the escape of superfluous moisture. That these fen lands are capable of producing abundant crops is evidenced by the extent to which the horticultural industry is carried on in the districts where soils of this character abound. The growers understand their soil and have learned by practical experience the best methods of treatment.

It is they who find chemical fertilisers of special value, and in many instances the strong manures, guanos, dried blood, bone-meal, hoof and horn shavings and other substances which are rich and, of little bulk prove to be of greater value than manures containing much matter of a humus forming character. On the other hand, the sand soils of the higher and drier levels of the coastal area demand much vegetable matter, and cool, moist manures such as cow manure are of greatest value, whilst clay may frequently be used to immense advantage, especially in seaside gardens on the east coast.

So far as first steps in cultivation are concerned they may thus be summarised: Disturb all soils to the greatest possible depth, expose as much of the soil as possible to winter's frosts, and endeavour to rectify excess of one element by application of that which is deficient. Remember that when endeavouring to improve a sandy, rocky, or chalky soil by the addition of loam, marl, or clay, several light dressings of well weathered soil will be more effective and beneficial than a single application of large quantities. For a beginning use at any rate the bulk of rich manures and fertilizers on the top spit, although it is never sufficient to begin cultivation by digging only one spit deep.

The after treatment of both subsoils and top

spit must be more precisely governed by individual conditions or circumstances.

For areas of moderate extent the spade and fork should be the implements used from the outset. This book is written for garden makers and cultivators, and does not venture upon the broader lines of soil treatment for farming and forestry. The plough is therefore an implement which will not generally be considered suitable for use. It may, nevertheless, be legitimate and proper when a garden of great extent is to be created on land which is in a very difficult state for handling with ordinary garden tools to begin by breaking up the surface with a plough. In that case the ploughing should be done in late summer and early autumn. If heavy soil bakes and hardens after ploughing it must be allowed to rest until rains soften the clods, but where light soil is workable immediately after ploughing the fork should be brought into service for the purpose of shaking out all weeds, roots, and rubbish, which must be thrown up in heaps and burnt. All this should be done before the subsoil is touched, for although many weeds will rot when buried more than a foot deep, some will live and grow, proving to be a greater nuisance than ever when they have obtained a hold in a loosened subsoil.

CHAPTER III

PROMPT MEASURES WITH GROUND INSECTS

QUITE a usual plan in writing books upon cultural matters is first to tell how to grow plants or crops to perfection and then follow on with instructions regarding the destruction of insect pests. This chapter is placed in the fore part rather than toward the end of the book for the purpose of emphasizing the necessity for cleaning the land at the very outset.

Whether a plot of land has been down to pasture, has grown rough native grasses, brambles and bushes, or has simply lain waste and unoccupied by aught but weeds, it is almost certain to be infested with insects, and never will so favourable an opportunity for their extermination occur as when the soil is first broken up, and is to lie bare, idle, and exposed.

Pastures are notorious for their infestation with wireworm, and the click-beetles, which produce succeeding generations. This is one reason why turf should be pared off rather than dug in, and when this turf is stacked the layers should be sprinkled with one or other of the compounds which act as soil fumigants in order to asphyxiate the pests in the stack.

The mere act of loosening the soil will imperil many of the insects, for birds will always swarm to newly dug land, and the insects which appear on the surface will be rapidly snapped up. Unfortunately wireworm will not long remain exposed on the surface of the soil, but will soon burrow into the clods, so that great as the assistance of birds may be, it is unlikely to effect a complete clearance, and soil fumigants must be brought into service.

There are one or two points that should be borne in mind when using any of the vaporising compounds.

First, let it be remembered that wireworms are agile and capable of quick movement. Upon the first approach of gaseous vapour the insects will endeavour to escape, and they are capable of descending into the fairly hard subsoil as well as through the loosened top spit. Another point is that the substances which are compounded to make a soil fumigant commence to generate vaporous gas as soon as they are brought in contact with damp earth, but excessively wet soil checks vaporisation. So long as the chemicals lie upon the earth, exposed to the atmosphere, the gas will rise and be dissipated in the air instead of permeating the soil, and furthermore, if scattered on the surface and merely raked in, the gas will not penetrate downward more than an inch or two, but will quickly find its way out of the soil. It follows, therefore, that if soil fumigants are simply distributed over the heads of wireworms their safety is not seriously threatened, for they will simply escape into the subsoil, will sojourn there awhile until the gas above them has disappeared, leaving the pests still alive and unharmed.

Obviously the prudent method is to use the soil fumigant during the process of trenching; to keep a supply at hand, in a barrel, box, or metal vessel to prevent absorption of dampness from the earth, and to spread little by little, covering immediately with the next shovelful of earth, instead of distributing a quantity over a wide area and leaving it exposed for hours. By first dressing the subsoil and then the top spit as it is thrown over, the wireworms find their retreat cut off, and whichever way they turn they come in contact with fumes which will destroy them. Whilst advising dressing the subsoil as well as the top spit it may be remarked that when buried more than two feet under the surface it is well nigh impossible for the vaporisation to take place properly. It is comforting, however, that only a few roots which wireworms are likely to injure will travel far beyond the two feet depth except it be parsnips grown for exhibition and some of the deep rooting herbaceous plants such as echinops, bocconias, crambes, etc., and ere these are established it is to be hoped the soil will be completely freed from so troublesome a pest.

There are three varieties of click-beetle and their larvae or worms, each of which is shown in our coloured plate. To the practical gardener all are equally undesirable, and their extermination is imperative. Gas lime is often recommended as an agent of destruction, and on bare, unoccupied land it may well be used. Whilst, however, gas lime does considerable good upon a stiff, tenacious clay, it is to a certain extent detrimental to very light, sandy soils which are deficient in humus, and the less violent ground fumigants, such as Gasonite, Vaporite, Kilogrub, Abol Soil Insecticide and various other proprietary compounds of a like character are preferable. A very good soil fumigant may be made by mixing common commercial naphthalene with lime. One pound of naphthalene thoroughly mixed with fourteen pounds of freshly slaked lime makes a powerful mixture, which, however, should be used quickly, and not left lying about to lose its strength. It is essential that the naphthalene shall be finely powdered and not used in flakes or hard lumps.

Leather-jacket grubs, the larvae of the crane fly or daddy long-legs, will be another likely pest where rough grass has occupied the ground. The crane-fly lays its eggs at the base of tufts of grass, and at the collars of plants of soft succulent growth, upon which the grubs will feed. Similar methods of destruction should be adopted to those advised for wireworm, and being slower of movement, and more prone to remain on the surface of

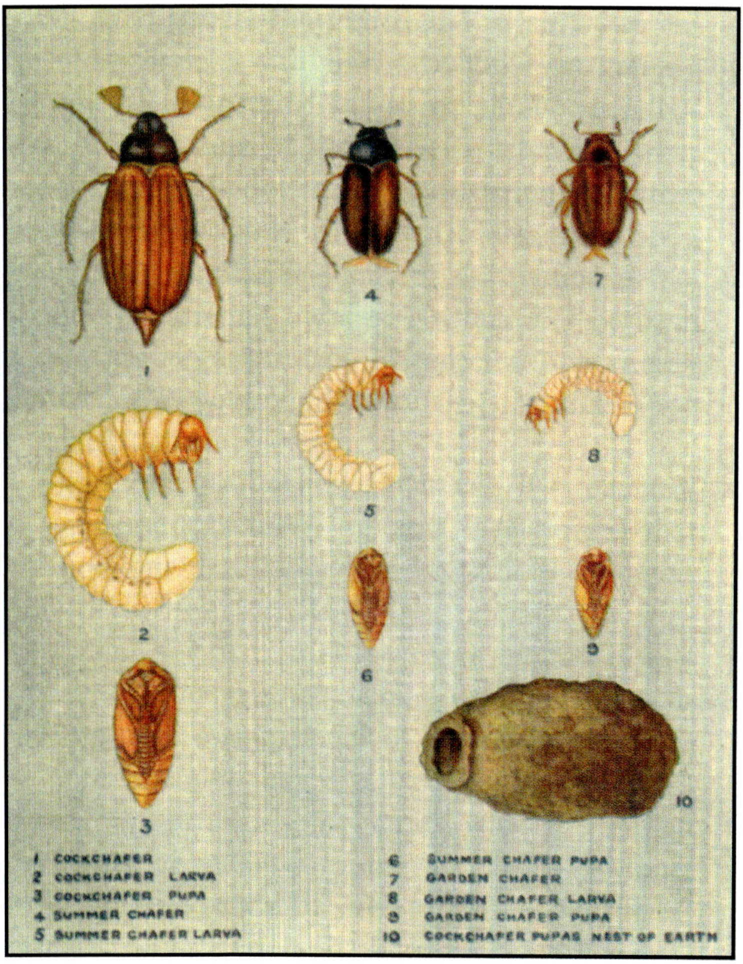

VORACIOUS GRUBS THAT FEED ON PLANT ROOTS

the soil, the birds and the chemical dressings alike have a greater chance of extinguishing these ravenous grubs. The chafer grubs, the millipedes, and many other insects may be found in soil that has lain more or less idle, and all should be given a hot time. During later stages of garden developmept, when plants of varying character are introduced there will doubtless be invasions of other insects and parasites which will demand either a repetition of similar methods of destruction or the adoption of different measures, which will be dealt with later, but as already stated, the opportunity afforded at the outset for cleansing the soil is too valuable to be neglected.

With every justification, doubtless, it is claimed: by some manufacturers of soil insecticides that they combine fertilising powers with their capability of killing insects. That statement is supported and verified in some cases by the results of chemical analysis, but whilst certain of the ingredients have manurial value, it is not to say that the more liberally they are used the better it will be for the soil. In some cases what are called ingredients of manurial value are forms of lime, and whilst lime may be and usually is beneficial to newly cultivated land, it is inaccurately described as a manure. Some of the substances used in manufacture of gas generating mixtures are of a character which it is undesirable should be introduced in excess to cultivated soil, and although it may be harmless to apply slightly larger quantities to vacant land than would be advisable where young growth and tender seedlings are concerned, it is still prudent to be content to follow the guidance of the manufacturers, whose instructions for use are naturally framed with a view to ensure a combination of safety and success.

There are certain chemical manures which insects dislike sufficiently to seek escape from; there are none that I am aware of which possess the power to kill these insects, but even to drive them away from particular plants may sometimes be an advantage. It is significant, however, that the kinds of fertilizers which are credited with such power over insects are usually kinds which should be fed to plants in only moderate quantities. Sulphate of ammonia, for instance, is sometimes claimed to be capable of ridding a plot of wireworm or leatherjacket, but as a fertilizer it is not desirable to use it too liberally or too frequently. Its first effect is to promote rapid development of foliage, and on that account it is a useful fertilizer for cabbage and other green crops, but not so good for hard wooded subjects or for flowers of substance. Used to excess, sulphate of ammonia actually has an impoverishing effect upon the soil, and wise cultivators make a practice of following a dressing of ammonia with a good supply of dung.

Ants in soil may become a great nuisance, and when anthills are discovered a hole four or five inches deep should be bored in the centre with a dibber, a dessert-spoonful of carbon bisulphide poured into the hole and the soil immediately closed over it. Keep carbon bisulphide away from fire, lighted cigarettes, etc., for its vapour is highly inflammable.

Neglect no opportunity for trapping moles on land which is to be converted into a garden. When their presence is revealed by the familiar mole-hills it will be well to catch as many as possible before digging operations are commenced, and likewise the nests and burrows of field mice, voles, rats and rabbits should be worked completely out at the earliest opportunity.

Mention is frequently made of what are termed "gardener's friends," which include various kinds of ground beetles, testacella slugs, centipedes, etc., and it is certainly well to refrain from crushing such useful devourers of enemy insects under foot whenever they may be observed. Unfortunately, however, the soil fumigants which will kill plant eating insects will also kill those which would prey upon the pests we cannot tolerate, and whilst this may be cause for regret, it would be folly to preserve them upon the mere uncertain chance that the friends will conquer and annihilate the foes. The very fact of the carnivorous insects being present in soil gives the cue that the vegetarian pests on which they feed are also there, and after all, if by fumigating the soil it may be cleared of the injurious insects, there will be less occasion to succour the friendly guests.

CHAPTER IV

DRAINAGE

HOWEVER well soil may be cultivated, the benefits of the labour will be largely discounted so long as defective drainage remains unremetlied.

It does not follow that every garden requires a system of pipe drainage, nor even that the laying of pipes will always prove to be beneficial, but it means that it is wise to ascertain at the outset what the conditions in regard to drainage may be, for neglect of this important matter is frequent cause of trouble and disappointing failure.

It must be quite obvious that gardens situated on high hillsides are less likely to be badly water-, logged than those which lie in hollows or on quite flat but low-lying ground. Elevation, however, does not invariably mean perfect drainage, nor is every low-lying garden troubled with stagnant moisture. The character of the subsoil has a deal to do with drainage, and in the ordinary course a solid clay soil is most likely to need drainage.

The points to bear in mind are these: growing plants like moisture so long as it passes quickly

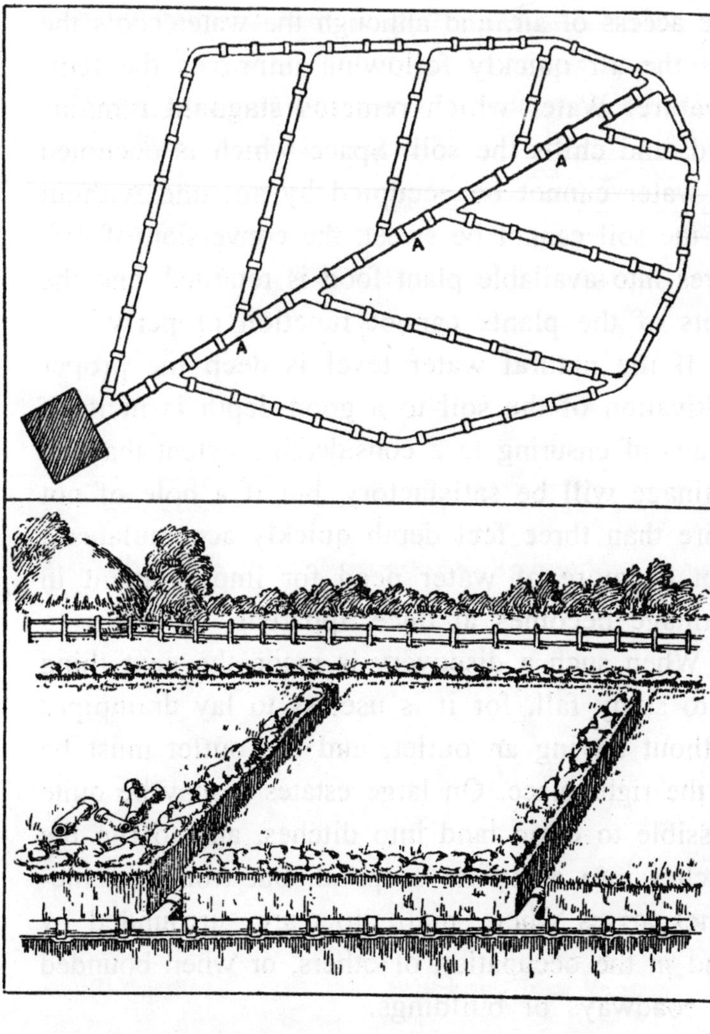

Top drawing-the general scheme of a system of pipe Driange.

AA-The main drain.

Bottom drawing- showing trenches for the pipes.

through the soil. The quick passage of water means free access of air, and although the water cools the soil the air quickly following improves the temperature. Water which remains stagnant remains cold, and chills the soil. Space which is occupied by water cannot be occupied by air, and without air the soil cannot be sweet, the conversion of manures into available plant food is retarded, and the roots of the plants cannot function properly.

If the natural water level is deep the proper cultivation of the soil to a good depth is in itself means of ensuring to a considerable extent that the drainage will be satisfactory, but if a hole of not more than three feet depth quickly accumulates a foot or more of water need for improvement in drainage becomes at once apparent.

When such a discovery is made the next thing is to study fall, for it is useless to lay drainpipes without having an outlet, and the outlet must be in the right place. On large estates it may be quite possible to drain land into ditches, and to run the ditches into a stream or pond, but trouble sometimes arises when small plots are surrounded by land in the occupation of others, or when bounded by roadways or buildings.

District authorities will not always allow garden drainage to be let into their drains and sewers, but where this can be arranged it is advisable that the connections shall be made by a properly qualified person.

Whatever may be the chosen outlet, the main drain should be laid with a good fall to it, and the side drains must also have a fall toward the main drain.

The exact position and course of the main drain must depend upon the individual circumstances of the land, its contour, and the position of the outlet.

Main drains should not bend or curve, but should run straight from starting point to outlet. Usually it is a good plan to run diagonally from corner to corner, assuming that the plot is more or less square or oblong. The side drains may run diagonally into the main drain, herring-bone fashion, entering it alternately rather than directly opposite. It is an advantage to connect the side drains with a boundary drain, as shown in our diagram. Nothing less than three-inch pipes should be used for the side drains with a four-inch or six-inch main drain according to the moderate or excessive quantity of water to be dealt with.

The actual planning of draining systems is, however, open to much variation, and the method will be adapted to meet particular conditions if the work is placed,. as it should be, in the hands of a practical workman who knows his job. The correct depth of the drain is a matter of importance. On heavy soils it is frequently necessary to have the drain at approximately two feet depth; on light sandy soils three feet may be more desirable or even four feet is at times a serviceable depth. Sometimes in draining a lawn pipes are necessary at a foot under the surface, and in very swampy places both deep and shallow drains are perhaps required to ensure really sound drainage.

Remember that the purpose of the drains is not to receive the water as it sinks from the surface during rains but to take up as it rises above the natural water level. Where that water level is too near the surface the drains need to be laid deep. enough to keep the water below that level, and in such circumstances there must be a limit to the area served by one main drain. Another point is that a drain is unable to draw water a great distance through a stiff clay, and the side drains in consequence must be fairly close together. At the same time it will be unnecessary for the diameter of the pipes to be as great in stiff soil, because the water will reach them only at a slow pace. On a sandy soil, in a wet locality, the water will travel quickly for a considerable distance to the pipes; the drains should consequently be of larger size, but wider apart.

The question of the advantage or otherwise of rubble pits and rubble drains is worthy of some attention. In small town gardens it may be a case of either leaving things as they are or of excavating a deep hole at the lowest point of the garden, filling it with broken bricks, stones and coarse, porous rubble, and although such a plan is but a poor makeshift for a proper drain, it is better than nothing, and is in fact of value in direct proportion to the depth and cubic capacity

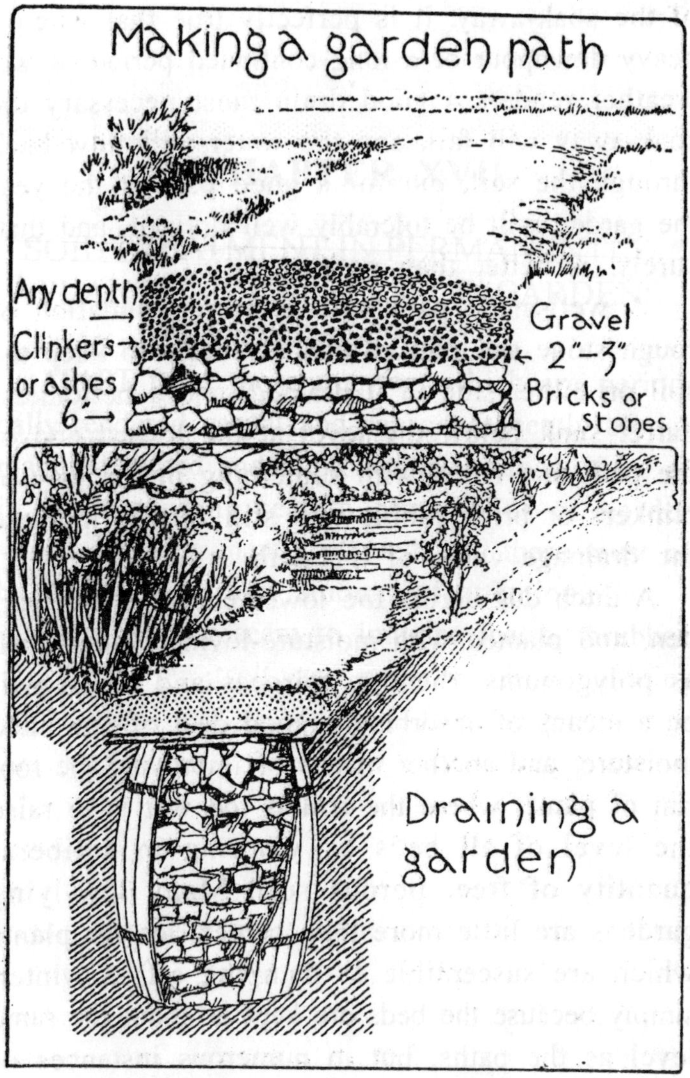

MAKING A GARDEN PATH

The rubble base of a garden path makes a useful drains, espacially when a soakaway is made as shown at the bottom of the sketch.

of the soak-away. It is perfectly true that when a heavy downpour or a long continued period of wet weather renders a good drain most necessary the soak-away will fail, and the water will flow back through the soil, but for a good part of the year the garden will be tolerably well drained, and that, surely, is better than nothing.

A well-made path with a deep foundation of rough stone and clinker will also help to keep the soil on either side of it drained, and a bottomless barrel sunk below the level at the lowest end of the path and filled with something in the way of clinkers or brick rubble will still further increase the drainage value of the path.

A ditch dug across the lower end of a wet garden, and planted with moisture-loving plants, such as polygonums, rheums, spiroeas and reeds, will be a means of absorbing a good deal of unwanted moisture; and another means of improving the root run of plants where the soil is too wet is to raise the level of all beds by working in a liberal quantity of free, porous soil. Many low-lying gardens are little more than cemeteries for plants which are susceptible to damping off in winter, simply because the beds are kept at about the same level as the paths, but in numerous instances of my acquaintance such gardens have been made tolerably comfortable homes for plants which have the reputation of being difficult, the method being to break up the garden into numerous small areas with dividing paths, and heaping up the soil almost in the manner of orthodox asparagus beds. Elevation of the crowns of fleshy plants naturally tends to keep them in a drier state than when they lie flat at a level which gives surface water no means of speedy travel. The use of porous stone chippings, clean shingle, sea-sand, burnt clay and charcoal on the surface of spongy wet soils is of material benefit, and although under ordinary circumstances trees with ramifying, ravenous roots are hindrances to good plant culture, there is always justification for planting a clump of strong growing shrubs or an isolated willow where the wetness of the soil is a constant source of trouble. The shrubby spirreas and some of the ornamental reeds and perennial grasses reduce the quantity of moisture in the soil to a considerable degree, and one may do a good deal of good to the rest of the garden by distributing clumps of astilbes, lythrums, lysimachias, Senecio clivorum and Siberian irises in positions where they can drain neighbouring beds which are required for other classes of plants.

CHAPTER V

DIGGING

THE reader may feel tempted to skip this chapter, under the impression that it is unnecessary to suffer the boredom of reading through pages of instructions upon so commonplace and simple a task as digging a plot of land. It is a great pity that digging should be so generally considered commonplace and simple, for such an impression is calculated to rob the task of interest, to encourage carelessness, and to prevent one learning to dig well. Until one fully realises the importance and the beneficial effects of digging there is no hope of effecting great improvement upon the soil, no hope indeed of becoming a successful grower. There is no soil so good that it cannot be improved by good digging; there is no soil that will not be spoiled if it is not well dug. Different kinds of soil call for different methods of treatment, but all require skilful and intelligent handling, including careful digging at the right time.

It is never the right time to dig land when it is saturated or when it is covered with snow. Even light, porous soils should be left alone at these

TRENCHING

The heap A is top soil, and B second spit from the first trench opened.

times, but the harm they suffer is small by comparison with the detrimental effect of digging heavy clay soil when it is excessively wet.

One cannot dig well when land is thoroughly dry; the proper condition is when just comfortably moist, but not wet enough to cling tenaciously to the spade and the boots.

No garden should be made without first digging the land at least two feet deep; three feet is better if the subsoil is movable to that depth. It may not be policy to dig so deeply every year, but even once is capable of considerably improving the soil for several years, for when once well broken up the subsoil will not settle quite so closely for a very long time. It is, however, highly desirable to disturb as great a depth as possible every year or second year.

Digging three feet deep is termed trenching; digging two feet deep is termed bastard trenching, and one spade's depth is termed plain digging.

The method of trenching may be explained as follows. First divide the plot to be dug into two equal parts by a longitudinal line through the centre. Next mark off across the end of one half-plot a strip three feet wide. Dig out the soil from this strip to a depth of one foot, cutting the sides of the trench thus formed evenly and perpendicularly. The soil removed from this trench should be wheeled across to the corresponding end of the second half-plot and cast up in a heap (A in accompanying diagram). Now measure one and a half feet breadth of the trench made and dig out the soil from it, transferring it to heap B. Having done this take a strong fork and break up the bottom of the narrow trench as deeply as the fork will go. To do this it will be necessary to stand in the trench and work backward. Into this bottom trench rough material may be worked, but it is unnecessary to use good manure, If the subsoil is clay, rough road grit, ashes, straw, grass, or tree leaves, or any wholesome humus-forming material that is rather too coarse for the top spit may be buried to open up the stiff subsoil. A sandy or gravelly subsoil will benefit by the addition of clean fresh cabbage leaves, annual weeds, rough bracken tops, or cut grass, but not roots of thistles, dandelions, etc. The great thing is to break the lumps and clods of soil as much as possible.

The soil from the other one and a half foot of the first trench is then turned over on to the loosened sub-soil, broken up as much as possible, adding more humus-forming substance, with grit, if the soil is heavy, and some long straw manure if such can be spared, the object being to improve and enrich this layer as much as possible.

Whilst lifting and throwing over the earth in this manner much loose soil will fall from the spade back into the new trench. All this loose soil must be scooped up with a shovel and evenly distributed over the dug portion, so that uniformity of depth may be ensured, which is only possible when trench by trench is made of equal depth and dealt with in a careful, methodical manner.

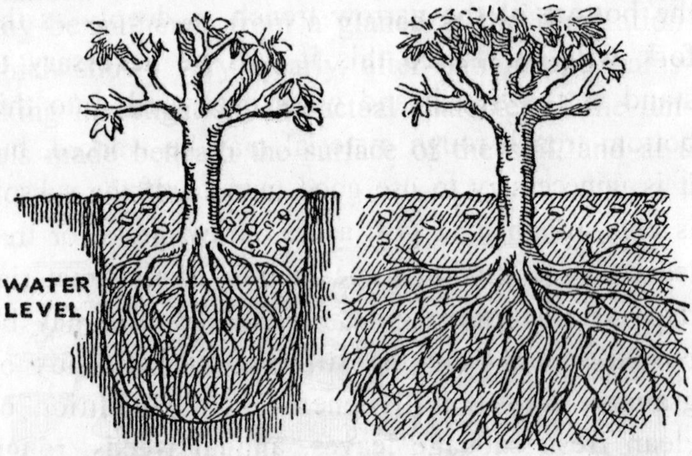

Left-shows a tree, badly planted in a hole made in hard, untrenched soil, forming a basin which becomes flooded in wet weather. The right hand drawing shows healthy root development of a tree planted in deeply trenched soil.

There is more importance in this matter of careful regulation of each section throughout than is frequently realised. Too often the idea is allowed to prevail that when time presses the subsoil may without serious detriment be skimped somewhat, the contention being that so long as the greater part of a trench is loosened and broken up it is of small account if a few ridges or the ends of trenches are left undisturbed. Such, however, is a grave error of judgment, for digging is one of the tasks which should be thoroughly well done, and may almost as well be left undone as half done. When some subsoil is loosened and ridges around are left solid the loose soil becomes puddled in wet weather, irregular drainage being brought about by the uneven density of the subsoil. Furthermore, when deep roots find hard resisting masses of soil they confine themselves to the looser parts, and the consequence is parts of the subsoil never perform any useful service in nourishing the plants. Roots may be twisted and contorted in all manner of shapes and still contrive to do their work, but they cannot do that work as comfortably or as well as they do when allowed to spread and develop in a perfectly natural manner.

Whatever may be subsequently done in the way of surface cultivation, soil enrichment, regulating the distance between plants, and so on, will be of benefit to a degree directly and inevitably dependent upon the quality of workmanship in the treatment of the subsoil as well as the top spit There is as much real science in the mechanical treatment and what is termed the manual labour bestowed upon a plot of land, as in the selection and application of chemicals calculated to correct the balance of properties in the soil. This should be welcome knowledge to all who own and cultivate a piece of land, but what is most wanted is a clear understanding of the fact that the fundamental, difference between the scientific and the unscientific is that the former is eager for accuracy down to the smallest detail, whilst the latter is prone to be content with "near enough."

By glancing at the diagram it will be seen that when trench 4 has been forked up and the soil from 5 has been placed over it, trench 6 can be broken up. The next step is to measure off another strip of surface, this time only one and a half feet wide. The soil from this (No.7) will go forward to No.1, thus bringing No.3 to the final level. No.8 can next be transferred to No.5

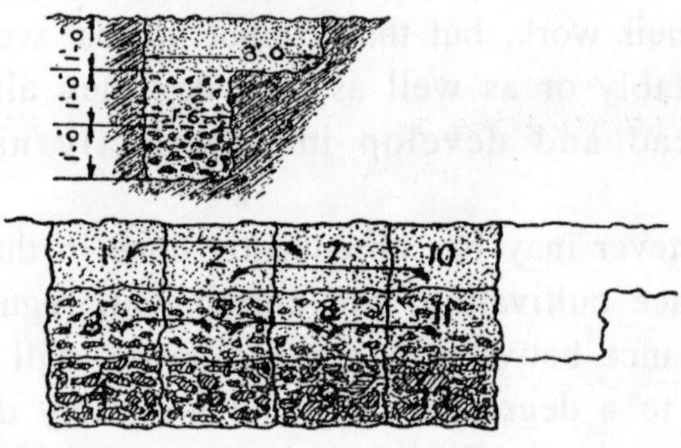

DEEP TRENCHING
Diagram of end view of trenched ground.

(over No.6), and that gives access to No. 9 for breaking up. No. 10 goes forward over No.5, No. II over No. 9, and so on until the end of the first half of the plot is reached with the top of the trench open three feet wide and the second spit one and a half feet. The soil to fill these vacancies must come from the second half of the plot, but to get at that which is to go in the one and a half feet trench the top soil must be heaped up close by, and again removed to its final position

Turning over Second Spit in 'Bastard' Trenching. (62a)

Cleaning Trench Preparatory to Turning Over Bottom Spit.(62b)

when the second spit has been transferred. Ultimately the soil taken from trenches 3, 2 find 1 fills in the last trenches of the second plot.

BASTARD TRENCHING is much simpler, and of course can be done much more quickly. On good well-drained soils it is equal to most ordinary needs but on stiff, hard subsoils the advantage is all in favour of the deeper trenching. For

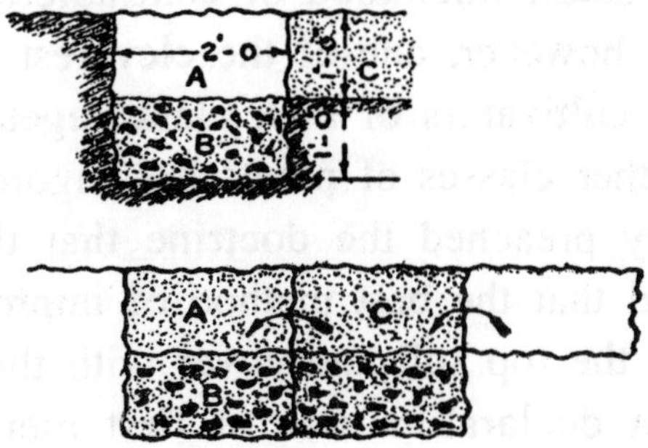

Top drawing-shows first trench A open, and second spit B, broken up ready for covering with soil from C, thus opening second trench.

bastard trenching one trench of top soil two feet wise is first removed, the soil below it being turned over. The next strip of top soil goes over to fill the first trench, the second trench thus opened being broken up and covered with the third strip of top soil, and so on throughout the plot.

It will be noted in both deep trenching and bastard trenching the method adopted ensures that each spit or layer of soil retains the same relative position after digging as before; in other words, the top soil remains at top; the bottom soil at bottom, and whenever the question is asked why such an elaborate business should be made of digging as has here been described, the reply will probably be given that it would ruin the land to bring the subsoil to the surface.

Until comparatively recent years one might have made that statement verbally or in part with extremely small likelihood of contradiction. For a long time, however, one of the cleverest and most successful cultivators of high class vegetables and of most other classes of plants has vigorously and persistently preached the doctrine that this is all wrong, and that the best method of improving soil is to bury the top soil and cover with the subsoil.

Such a declaration was at first met with be wildered doubt, by questioning as to whether the pen had not played tricks with the intentions of the writer, and there were not wanting those who vehemently denounced the heresy of this strange doctrine and held it up to scorn and ridicule.

The teacher of the new faith, however, had the courage of his convictions, and persisted in reiterating his advice, and because of his unquestioned knowledge and ability, coupled with his gift for lucid and forcible expression of opinion, a certain number of lesser lights, writers chiefly of smaller practical experience than of journalistic skill, took up the same theme, and to-day one may read that the old notion is a fallacy and this the writers set to work to prove by certain more or less emphatic statements.

To unconditionally repudiate this new teaching is to disclose blind prejudice and a disregard for certain stubborn facts, but to declare or to assume that it is in any way reasonable to make a regular practice of reversing the positions of top and bottom spits is to display equal disregard for other facts which are quite as plain and as important.

The superiority or greater fertility of the top spit of soil is largely due to its position, close contact with the atmosphere and its sweetening influence, accessibility for the purpose of frequent disturbance with spade, fork, or hoe, and favourable opportunity for receiving deposits of various substances of manurial character, all being factors in the matter of soil improvement which the subsoil lacks, to its serious detriment. These statements must necessarily support the view that subsoil will improve if brought to the surface, exposed to the atmosphere, dug, hoed, manured, and it undoubtedly will do so. Coupled with this fact is the benefit the roots of plants and trees will reap from good, sweet, fertile soil placed down where otherwise there would have been subsoil which would have only acidity, inertness and poverty to offer the roots of whatever may be planted in it. Here, then, is the case for the advocate of reversing the positions of top and bottom soil; but it is prudent to consider the arguments of the opposition, for there are very strong points in these arguments.

There is no advantage in questioning the statement that sub-soil which has long lain buried, undisturbed, and tightly compressed is unfertile, acid, and in a totally unfit condition for immediately sustaining plants or encouraging healthy growth.

Such soil needs long exposure to the atmosphere to be made sweet, and as every heavy downpour of rain which falls upon soil in that mechanical condition will consolidate it and shut out the air it will require to be very frequently forked, hoed, or otherwise broken up to readmit air. It will require nourishment and humus, and most probably a certain amount of lime for the purpose of correcting acidity, and at best a period of some months will be occupied in bringing the soil into a sufficiently fertile condition for even rough vegetation. Whilst, therefore, the good soil below is capable of nourishing plants which have large enough roots to reach and work in the lower strata, the infertile soil on the surface is incapable of promoting healthy growth of tiny seedlings and small young plants until it has been mellowed and sweetened by the atmosphere and the various methods of cultivation. One other point must be borne in mind: the good soil which is buried beneath the sour unhealthy soil will receive a certain proportion of the impurities of the latter as they are loosened and broken up by atmospheric

LEVELLING UNEVEN GROUND FOR MAKING A LAWN

Top-driving levelling pegs.

Centre-Top spit heaped around the plot, subsoil levelled.

Bottom-Top soil evenly distributed over level base.

influence and washed down by rain. Lack of air will also have its influence upon the buried soil, and the longer it remains imprisoned and unoccupied by active roots the more it will lose of its sweetness and fertility.

Skilful handling, industrious labour, and liberal applications of suitable fertilizers and humus forming products will bring moderately bad subsoil into a tolerably fertile condition within a year, but during that year nothing of a choice character in the way of small plants can well be grown. Trees, shrubs, and large deep-rooting subjects will be safe enough, so long as the surface is kept open by frequent disturbance and is improved in texture and quality, and in view of the fact that deep trenching cannot well be performed after permanent subjects such as trees and shrubs are planted, the burying of top spit soil before planting has advantages in connection with the creation of orchards, shrubberies and tree plantations. For flower gardens, vegetable quarters, and for lawns the older and generally adopted method of keeping the soil strata in their original positions must certainly be considered the safer course for the cultivator of limited experience and for those who desire to form and furnish a garden with the least possible delay and moderate expense. Let it be borne in mind that the mere fact of breaking up the subsoil will tend to effect some improvement, which may be further increased by working into the bottom trenches humus-forming substances, grit, ashes, lime, etc., according to their suitability for the particular character of the soil under treatment.

Whenever opportunity offers it will be highly beneficial to retrench the ground, again adding improving agents to the subsoil, and after awhile the lower strata will be so much improved that it may quite reasonably change places with the top spit, preferably choosing a time when it will have a winter's exposure in the rough before being raked down and planted.

Special precautions are necessary in making ready a site for a lawn, especially when the land is originally very uneven in surface. The top soil, which is invariably the best, must be kept at the top; therefore before removing hillocks into hollows the whole of the top soil to a spade's depth should be taken off and wheeled to heaps conveniently near the site. Then with the customary methods the undersoil should be brought to a level, and the good top soil, together with any enriching materials, such as hoof and horn shavings, bone-meal, etc., be evenly spread over the whole area. This is the only satisfactory means of ensuring a sound. base which will provide fine lawn grasses with a congenial rooting medium in which uniform growth can be produced. The uneven, patchy appearance of a vast number of lawns is due in the majority of cases to carelessness in burying the proper surface soil in the hollows, leaving only inert sub-soil elsewhere into which the grass cannot root with equal readiness.

SURFACE DIGGING

is quite a different matter from deep and batard trenching. It is essential at frequent intervals and for various purposes, and not least among them is that of disturbing and destroying both seedling weeds and the larvae and pupae of injurious insects. Surface digging is the means of opening the soil to the atmosphere and of preparing the ground for reception of seeds and small young plants. It affords opportunity for incorporating manures and fertilizers, sand, wood ash, and other improving agents as circumstances may require, and if properly carried out at every convenient opportunity, it may be the means of considerably increasing the depth of sweet and fertile soil. These remarks may appear to be somewhat in the way of stressing the obvious, but they are made simply because it is too frequently apparent that the only ideas in the mind of the cultivator when surface digging are of facilitating easy planting and of burying manure. The danger is that in learning of the benefits of manures and fertilizers the cultivator may be led to assume that chemical treatment of the soil is the one and all important consideration, whereas in truth the mechanical treatment is of equal, if, indeed, not greater importance.

Much depends upon the choice of proper time for digging, and the condition of the soil should always be the deciding factor in that matter. It is very frequently asserted that heavy clay soil should be dug in autumn, and that light soil may be dug at almost any time. As a general rule there may be little wrong with this view, but it would be infinitely better for land cultivation if general rules were not accepted too readily as being invariably correct, for the truth is soils vary so much in character that the best treatment can only be given when individual circumstances are very carefully considered.

A very tenacious clay dug in early autumn may begin at once to disintegrate, but before even the surface of the clods has become well loosened. heavy autumn rains may convert the whole once more into a plastic paste, and when that happens by October or November there is little likelihood of disintegration proceeding further. December and January frequently prove the better months for surface digging very stiff, heavy soils, keener frosts accomplishing their beneficent task to fullest advantage.

No soil should be dug when so saturated with moisture that it clings to the spade and to the boots. That simple statement is so threadbare with usage that one doubts the effectiveness of reiteration. Apparently the idea lurks in the mind that the main reason for adopting this ban upon digging wet soil is an effort to study the comfort of the digger, and therefore when seasons are wet and much labour lies ahead, discomfort is ignored and digging proceeds, though the soil hangs like putty to the spade and the surface churns up like porridge, clinging to boots in sticky masses. One may admire the dogged determination of the digger, but must deplore the serious harm suffered by the soil. There are, it is true, some extremely porous soils that contain so little of either clay, chalk or humus that they very rarely hold sufficient moisture to either cling or become solidly compressed, but whoever possesses such soil, despite the fact that it may be dug immediately after a rain, should make strenuous efforts to strengthen and improve its texture, for hungry as it must be, it is almost futile waste to put good manure and fertilizer into soil that has no absorptive power. There are some soils which would be accounted quite light when normally dry which when saturated are as unworkable as clay, and when thoroughly dry are impenetrable bricks. Early autumn is the safest time to dig such soil, because postponement may mean loss of the solitary opportunity for good work, and if later on it should be found possible to turn over a second time, so much the better for the land. "Thought attempting to enter into the scientific explanation of the process by which effect is produced, it may suffice to say that the trampling and moving of saturated soil upsets its porosity and friability, squeezes the particles together to the exclusion of air, and results in hard caking when eventually the moisture dries out.

Far too much, surface digging is done with a fork, and that is not the implement which facilitates workmanship of high quality. For one thing, it is difficult to maintain a good level surface with a fork, but it is easy to break down the soil to too fine a tilth. Roots of weeds too readily pass through between the prongs of a fork back into the trench to be hidden by loose soil which follows in their wake. A spade should be used, and in autumn and winter the clods should be cut and turned over evenly to be broken down by atmospheric action, but in spring it is better to break down the clods as digging proceeds, lest a spell of dry weath results in baking the soil like bricks.

There is something in the sight of a plot of land evenly dug in unbroken clods which bespeaks good workmanship, but it is not for appearance alone that land should be thus dug. Our first diagram of plain digging shows the appearance of the clods immediately after digging, but the second shows how the same plot will have crumbled down and changed the whole appearance of the surface after exposure to a winter's frosts. There is nothing that will so effectively and beneficially work this change as frost, and the soil to benefit most is stiff, stubborn clay; hence the advisability of timing the digging of clay soil to secure the maximum benefit from frosts before heavy rains have battered down the surface and filled in the

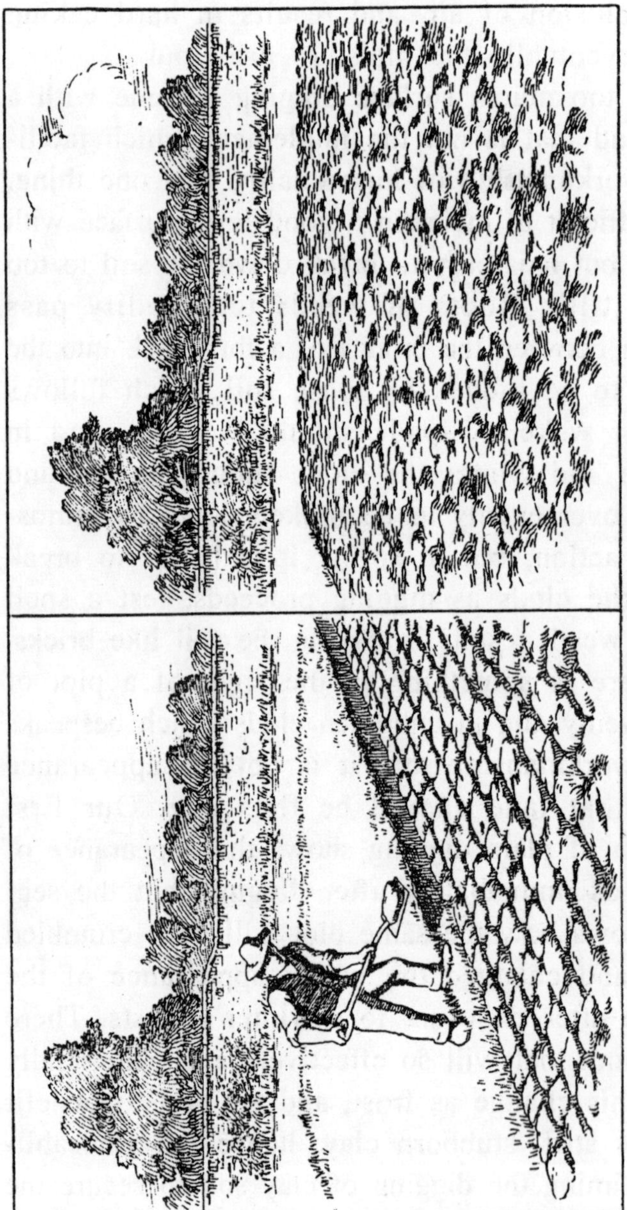

PLAIN DIGGING HEAVY CLAY SOIL

Left-Turning over in clodes, leaving in the rough. Right-The surface, showing clods disintegrated and crumbling after winter's frost. Arake will now produce a fine tilth.

crevices with pasty, muddy soil. To dig properly, a trench one and a half feet wide and one foot deep should be opened, and thereafter, working in regular, even rows, each spit should be turned completely over, and thrown to rest in its position to maintain an even level of the dug soil.

One of the purposes of surface digging, as already explained, is to incorporate manure and fertilizers. There is a right and a wrong way of doing this; the wrong way is to throw into the bottom of the trench solid masses of fertilizer and cover these with solid masses of soil. The right way is to evenly distribute the manure so that the clods are equally surrounded by manure except on their surface. The reasons will be made apparent in due course.

CHAPTER VI

CORRECTION OF MECHANICAL INEQUALITIES IN SOILS

APART from trenching, digging, draining and even manuring, the majority of soils require other measures to secure a proper balance of constituent parts, and in a very large number of cases treatment in this particular respect is of an importance fully equal to that of any of the other requirements. Few soils are really perfect in constitution and texture, whilst those which by long cultivation are most improved in texture frequently suffer from acidity or sourness due to a prolonged period of liberal manuring.

CLAY SOILS

involve the greatest amount of hard labour, but although originally totally unfitted for the cultivation of most plants and crops, a clay soil will frequently yield the finest results in response to judicious treatment.

The stubborn tenacity of clay is largely due to the close compression of minute particles of finely pulverised slaty or marble rock, of which it is principally formed. The addition of sharp sand, flint, granite or limestone grit, coal or coke ashes, will improve porosity to an extent dependent upon the even incorporation of the hard, gritty materials, but although this improvement of porosity is much required and highly beneficial, it does not actually in itself add to the fertility or richness of the soil. A very good plan is to dig out some of the stiff, dense subsoil and burn this on smother fires distributed conveniently over the plot. A number of small fires are always better for this purpose than one of great size, because in a very large fire with a great volume of heat maintained for a long time the clay gets over burnt and loses every particle of virtue, whereas when lightly burnt in a small fire and cooled within a moderate time the clay not only loses its tenacity but a small amount of hitherto unavailable plant food is liberated and rendered soluble.

Humus is deficient in a stubborn uncultivated clay, and must be provided in one or other of several possible ways.

Stable manure, rotted leaves, spent hops, lawn mowings will all provide humus, but Instead of burying these in dense masses they, like all manurial substances, should be spread and dug in, always intermixing the vegetable matter as thoroughly and evenly as possible. All the benefit of amelioration by the aid of humus will be nullified if a stiff tenacious clay is dug and trampled about whilst excessively wet, and it will be equal to acquisition of much scientific learning if the cultivator gains an intimate understanding of the condition in which his soil may be worked. Some seasons are very unfavourable to the successful handling of clay soils. The year 1925 affords an instance of the difficulties that may be confronted. The winter had been mild and moist with no frosts to disintegrate stiff clay, and the spring was so persistently wet that prudent cultivators hesitated to disturb heavy land, hoping that a spell of fine weather would improve its condition. Drenching rains were, however, experienced until the very end of spring, when suddenly drying winds arose, and instead of gradually improving the condition of the clods of clay the wind hardened their surface. Then followed a prolonged drought, with scorching sun, and as a result stiff clay soils in many districts remained unworkable right through the month of July, and many big areas of heavy land the writer saw which produced no crop throughout the season.

When one is fortunate enough to. have the option of selecting a site for a garden, a nursery or a farm, it is prudent to avoid extremely heavy clay, for not only does it involve a great deal of heavy labour, but there is frequently the disadvantage of necessary work being upheld through unsuitable conditions. Sometimes, however, it is a case of working a clay soil or none at all, and then the only thing is to make the best of things by putting as much work as possible into it when conditions are favourable, taking comfort from

Arum Lilies in the Open.(78)

the fact that skilful treatment of stiff clay will in time produce a soil that will yield wonderfully good results, growing many things a great deal better than very light soil, retaining moisture and nourishment for a long period and showing fullest benefit for every measure of improvement.

Too often the benefit derived from one season's good treatment is destroyed by injudicious treatment the next, for although a winter's frosts and a dressing of light material may bring a very stiff clay to a condition of nice friability, it may very easily be again reduced to a livery, tenacious mass by trampling or disturbing it while too wet. An example came to my notice recently in the eastern counties, where a field which had been long under cultivation and had been considerably improved was seriously harmed by being ploughed while so wet that the furrows cut, as the ploughman said, "just like liver." When the surface dried the whole field was like kilned bricks, and a whole season was lost, as well as the improved condition of the soil that had been the outcome of years of good husbandry.

An old-fashioned agricultural idea has been that once every several years a field should lie fallow, to rest it, and to enable perennial weeds to be destroyed. The fact is a clay soil may very easily suffer much harm through lying idle. There are bound to be periods during summer when the idle soil cannot be surface-cultivated, but when weeds will flourish apace, and it frequently happens that at the end of a season's fallow the weeds are more rampant than ever. An infinitely better plan when a plot of ground needs a rest from customary crops or requires cleansing is to sow it with tares, rape or mustard. With a quick-growing crop that makes abundant leaf, thickly covering the ground, weeds have little chance of making much growth. When the crop is well up, but before it gets too far advanced, it should be turned in, but it is highly necessary in the case of heavy soil that an opportunity should be taken to turn in the green crop when the soil is not wet. The process of green manuring is of far reaching importance, and is, in fact, deemed worthy of treatment in a separate chapter; it may well, however, be mentioned here as of particular benefit to clay soils and as being a splendid alternative to the sometimes unsatisfactory and never economical practice of summer fallowing.

In garden plots, of course, the green crop should be dug in with the spade, and during the process perennial weeds and their roots will be carefully picked out. Fields, of course, have to be ploughed and the weeds can only be dealt with by harrowing, but the drawback is that in harrowing out the weeds much of the green crop is also unearthed. On this account one of two methods has to be adopted, either harrowing out the weeds as thoroughly as possible before sowing the green crop or allowing an interval to elapse before harrowing after ploughing in the crop. This book, however, is not intended to cover the agricultural treatment of land where that treatment necessarily differs widely from the methods of gardening, this latter providing more than ample scope for present purposes.

To say that incorporation of sand is beneficial to clay soil is not quite enough, for there are sands which tend to bind rather than assist porosity, and it is highly essential when going ,to the trouble of sanding to carefully select the coarse, sharp sand that will never bind, but that will affect to the greatest possible extent the porosity of the staple soil. Sea-sand, river sand, shell gravel, and, for very stubborn clays, small shingle may be used with great advantage, and it is often of immense assistance to apply a dressing of such sand to the surface in addition. to any that is buried during the course of digging.

The reason for the surface dressing is that it will greatly assist the production of a fine tilth when hoeing, and that is of vital importance, because clay soils which are left without a fine loose tilth are bound to crack during dry weather, and that means escape of much needed moisture. Mulching during summer with chaff, grass, brewers' grains, spent hops or other loose litter is beneficial for the same reason.

LIMING CLAY SOILS.

The whole question of liming soils is of extreme importance, and although I often find myself at variance with the opinions of those who apparently consider it impossible to err in application of lime to soil, it may certainly be claimed that stiff clay soils that have not been long or well cultivated are almost invariably to be benefited by liming.

The purely mechanical action of lime upon tenacious clay is of no small significance, for a lime-dressed plot of stiff unwieldy clods will break down in a manner similar to that brought about by alternate freezing and thawing. The pulverization is really due to expansion and contraction caused by absorption and evaporation of moisture.

Clay expands as it absorbs moisture and contracts as the moisture dries out. Lime changes but little in bulk, whether wet or dry, especially when it is in a finely pulverised form. Thus when particles of lime are washed by rain into clods of clay they do not shrink during the drying out process to the same extent or at the same pace as the particles of clay, and the inequality of shrinkage causes the separation of the mass of clay. My only reason for explaining this point is to make clear one or two facts about the application of lime for the particular purpose at present under consideration.

The lime requires to be finely pulverised, because

that is the form in which it can most readily gain access to the dense masses of clay. Quicklime is most effective because it will take up a very large quantity of water which would otherwise go into the soil, and it generates great heat when the water is absorbed, hastening thereby the drying of the clay. Quicklime left actually on the surface, exposed to the atmosphere will become slaked, and the heat, rising into the atmosphere, will be wasted so far as benefit to the soil is concerned.

When just covered with soil rain will wash down particles of lime and the soil beneath will thus be brought into beneficial contact with the lime, but the deeper it is buried in the first place the less marked will its effect upon the tenacity of the clay become, for the simple reason that the pace at which deeply-buried clay can lose its moisture is much slower than that which is nearer to the surface and atmosphere.

There are other functions lime has to perform which render its use on subsoil beneficial, but for this particular purpose of pulverising the stiff stubborn clay it may certainly be urged that where the available supply of lime is limited and the need for speedy improvement is great it is prudent to confine the dressing to the top spit, but to get a slight covering of soil over the lime with as little delay as possible. Rest content that a good proportion of the lime buried in the top spit will soon be washed down to the subsoil, if the land has been well and deeply broken up, and that the subsoil will be benefited in consequence, but the greatest and speediest effects will be felt in the top spit, and that is where they are most urgently needed.

Slaked lime may be used instead of quicklime, but with slightly less conspicuous results. Ground limestone may be used, but here again action is slower than in the case of quicklime. Machine ground quicklime is far more effective. Chalk is of smaller benefit to clay soil, and should rather be used on light sandy soils that are found to be deficient in lime. It remains to be added that quicklime, although recommended for use on freshly dug clay soils, is too caustic for use among growing plants.

Another purpose for which lime is useful on a stiff inert clay is that it has the power to liberate chemical properties which are frequently present in the soil but in a form which prevents plant roots taking advantage of them. Potash, for instance, is usually abundant in clay, but only contact with lime will convert the potash into a soluble or available form for the nourishment of plants. Lime is a great soil sweetener, a disinfectant, and although not in itself a manure is of immense benefit as a component part of the elements of plant foods existing in the soil. These points, however, are more appropriately dealt with in the section devoted to fertilizers, at present our concern being the mechanical improvement of an untractable clay. Whenever plain digging is proceeding affords an excellent opportunity for application of lime, and it is more beneficial to give a light dressing at the time of autumn digging and another in spring (before planting) than to give the whole amount at one time. One pound per square yard should be the maximum quantity for a single dressing.

Gas lime may under certain circumstances be used on excessively heavy clays; on broken up pastures which are badly infested with ground pests, but should usually be confined to soils requiring drastic treatment and which are to undergo a lengthy period of preparation before any planting is to be commenced.

Gas lime contains a large percentage of sulphide of lime and also some sulphite of lime. These are both rank poisons to vegetation, and owing to their presence gas lime must be spread on the surface of the ground early in autumn and left exposed until spring, by which time the atmosphere will convert the poisons into gypsum, which is of fertilizing value, and may be dug in with advantage.

Noxious weeds as well as insects succumb to a dressing of gas lime, so that under certain circumstances this is a valuable material for agricultural and horticultural use. The necessary conditions are that the soil shall be in bad condition, necessitating a lengthy period of treatment, and that it shall be completely free from growing trees, plants or crops for a period of at least six months after the gas lime is spread.

CHAPTER VII

SANDY SOILS

IN violent contrast to the clay soils previously dealt with the most unsatisfactory and unprofitable are the loose, light, impoverished sandy soils which are never firm, are extremely hot in summer, very cold in winter, are incapable of holding moisture, and are as wasteful of plant food as of plain water. Even such soils should be deeply dug, especially as there will frequently be found somewhere beneath the surface a "pan" or hardened layer of fine particles of impurities, through which no roots will penetrate, whilst during periods of drought the moisture from below this pan is prevented rising by the usual capillary attraction, thus adding sorely to the distress of the parched plants. The most urgent need is to get humus into the soil. It is of very little use to apply rich, easily soluble fertilizers, because the first rains will dissolve and wash them away into the drainage. The addition of humus will provide a sponge which will soak up the moisture and any dissolved plant foods it may contain, and will hold the nourishment at disposal of the plant roots.

There are many ways of creating or providing

humus in such soil. Instances are on record where utterly barren, shifting sands have in the course of a number of years been converted into fertile soils by means of successive crops of sandloving weeds which have been turned in while in full herbage and followed by other crops. Starting with small sand-grasses, sufficient humus has been created to sustain a crop of common annual lupines. This has been followed by bigger crops and by vetches, rye and other green crops, until the humus has increased to an extent that made it worth while dressing with manure. Such are extreme cases, but on a modified scale the sandy soils which may be met in some of the flat districts of our own country and especially within a narrow margin of many parts of our coast line require similar treatment.

There are, let it be understood, some so-called black sands, as, for instance, those surrounding the Wash in the Fen district, and again on the low-lying marshes of the Dee and Severn, the Solway Firth and other places where the natural sand is already so largely impregnated with humus that has been washed over and deposited by flood tides that they have an excess rather than a deficiency of humus, but for the moment these are left out of consideration, and we are dealing with sandy soils which by their light colour, looseness and excessive porosity are readily recognisable as soils which sadly lack humus. They differ even from the stony red or yellow gravels, for these latter frequently, if not generally, rest on beds of solid chalk, although the gravel coatings are sometimes so thick that the presence of chalk below is unsuspected. In addition to green manuring by the method of growing crops for digging in, all manner of vegetable matter may be carted on to the ground and dug in. For seaside districts seaweed may frequently serve a good purpose. Where breweries are situated conveniently spent hops are to be highly prized, and many other materials will be pressed into service. The great thing is to evenly incorporate the humus-forming material with the staple soil; not to be content to fill trenches with solid masses and cover with other thick masses of the soil. Sandwich layers of this character are not desirable, for the simple reason that neither one substance nor the other is of itself a good medium through which tender young roots may have to percolate, a well balanced compost of any kind being only obtainable by thorough incorporation of the several ingredients.

The addition of clay to sandy soils is frequently advised, and especially in books upon rose culture. Very good and sound is the advice to add clay to sandy soil, but it should be supplemented by a few remarks regarding the kind of clay to be used and the method of its application.

Many roses have been done to death by being stuck into holes with great lumps of stiff, inert, unfertile clay rammed and trampled over and around their roots. In due time the moisture in the clay has been drawn off by the porous soil around, and the clay has set like lumps of rock during the first drought. When subsequent rains fall the moisture takes the course of least resistance and filters through the loose sand, leaving the solid lumps of clay still hard, while the roots of the roses make futile efforts to spread themselves and gain nourishment.

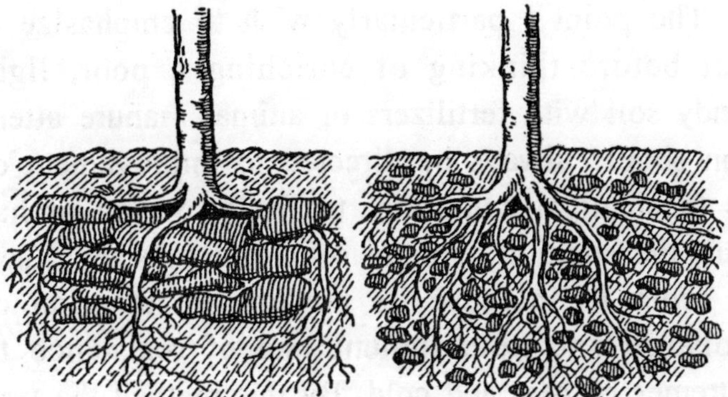

Left hand drawing shows lumps of clay packed round roots. This is wrong. Right hand shows clay evenly distributed through sandy soil. Such incorporation is highly beneficial.

Clay that is to be used for the improvement of sandy soils should itself have been brought to a state of fertility by good cultivation. It must be top spit clay, not sour lumps dug up from inert subsoil. It should be spread in a thin even layer over the surface of the ground in autumn, or winter, left exposed to the sweetening and pulverising influence of frost and air, and then, when in a nice friable and workable condition in spring, should be carefully dug in, mixing as evenly and thoroughly as possible with the soil. A very sandy soil cannot be immensely improved in one operation or in a single season, and permanent advantage is altogether on the side of successive applications of the improving agents in moderate quantities rather than a heavy dressing at anyone time.

The point I particularly wish to emphasize is that before thinking of enriching a poor, light sandy soil with fertilizers or animal manure attention should always be directed to improvement of its mechanical condition, to the provision of humus, the addition of more solid and substantial soil in order to increase its powers of retention of moisture and nourishment and of resistance to extremes of heat and cold. By this method the way is prepared for preservation of the highly desirable plant foods which manures and fertilizers will provide, whereas the alternative method of beginning with the application of nutriment is extravagant and to a large extent futile.

THE QUESTION OF LIMING SANDY SOILS

Careful analysis or even rough and ready tests will frequently reveal serious deficiency of lime in sandy soils. That deficiency requires to be rectified, but care in the method of procedure is very necessary.

Caustic lime applied to a sandy soil is capable of doing a great deal of harm. Quicklime when moistened burns, and as it cannot burn sand, which is simply stone in tiny particles, it will burn whatever vegetable matter or humus is with in its reach. It would be folly to laboriously add to the humus in sandy soil and then add quicklime to burn it up. Ground limestone is less detrimental, but chalk is the safest and most suitable substance to use on sandy soil, and whereas quicklime tends to break down and pulverize stubborn clays. Chalk tends to bind together particles of sand, thus still further improving the moisture-retaining power of the soil. In regard to a suitable quantity to apply, the cool, mild character of chalk renders it safe to use in liberal quantities, a half-hundred weight to the square rod is not too much to use where lime is known to be deficient.

When supplies are obtained direct from chalk pits exposure to air will be desirable before the dressing is dug in. Hard lumps will crumble down in the course of a few weeks if they lie evenly distributed over the surface of the soil, and, unlike caustic lime, chalk undergoes no chemical change as a result of absorption of moisture or atmospheric gases.

A good plan, therefore, is to spread the chalk over the surface as autumn digging proceeds, leaving it thus until spring.

CHAPTER VIII

STONY GRAVEL

To a certain degree the shortcomings of stony gravel soils are akin to those of loose, light sand, but there are important points of difference. The sand in gravel is usually intermixed with clay and with chemical matter of a somewhat sticky, adhesive character. Gravel usually binds as sand of a sharper nature cannot do, and were it not for the great quantity of stones in various sizes and shapes, gravel soil would set so close and hard that ceration, percolation of water, and root action would all be very severely handicapped. The stone in gravel, although a very frequent cause of complaint on the part of cultivators, may properly be termed the saving element which renders cultivation possible, not only by reason of its assisting ceration and drainage, but because the stones distributed throughout the soil check the ascent of moisture from below in time of drought, preventing rapid evaporation.

In soils which have settled after digging minute hair-like tubes form themselves in vast numbers more or less perpendicularly from the surface downwards, and through these capillary tubes the moisture finds its way back from the subsoil to which it drained after rain, and passes out from the soil into the atmosphere, but when a stone intercepts the course the progress of the moisture toward the surface is arrested, and the whole area covered by the stone is slightly moistened before ascent recommences through

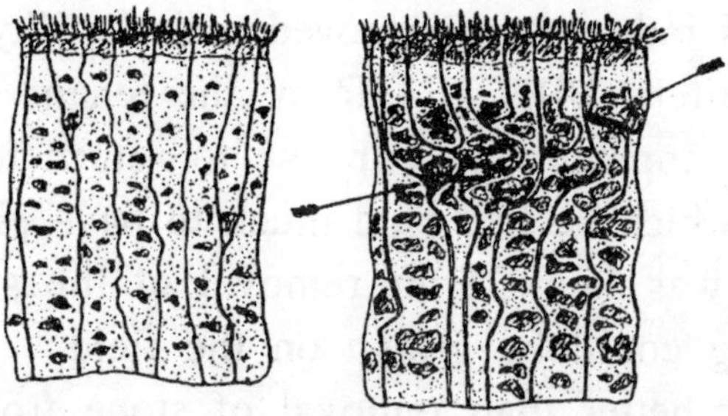

Left-conveys an idea of how capillary tubes may run uninterruptedly through comparatively stoneless soil. The tubes are actually much finer and more crowded than can be represented in a drawing. Right-shows how large stones divert capillary tubes, arresting rising moisture. The arrows indicate damp patches under the stones, which in varying degree will be found under all stones.

other tubes surrounding the edges of the stone. It would be impossible to draw capillary tubes, for they are far too minute, but the reader will grasp an idea of what happens by looking at the sketch which shows what may be taken to represent a few capillary tubes intercepted by stones. The darkened patches under the stones represent the moistened area to be found under a stone even In the hottest weather, and where stones lie upon the surface their removal shows the moist patches qlltte plainly. When, therefore, the advice is given that stones should be removed from the surface of the soil let it be remembered that drought resistance will be somewhat reduced and that the tendency to consolidate and set hard to the exclusion of air will be increased. Cases are known where the removal of stones from gravel soils has destroyed rather than improved their fertility, and on the Chitern Hills is a farm, the tenant of which had at considerable expense to return to a field stone which he had paid much to have picked off, for it was found after removal of the stone that nothing could be grown on the field.

Far better than removal of stone from gravel soils, except of course large flints, etc., which interfere with spade work and planting, is to increase the bulk of earthy matter, to trench deeply, whatever the subsoil may be, and to decide instead of persisting in growing such things as cannot properly thrive on stony ground to be content to concentrate upon those that can be well grown.

Deep trenching is often spoken of as useless where the subsoil is chiefly composed of big coarse stones, and still more so when the gravel lies only a foot or eighteen inches thick over solid chalk; but the fact is in either case the benefit accruing is immense, and where the solid chalk has been deeply broken up harsh gravel soils have by the aid of judicious manuring proved to be not only

Crinum Powellii of Exceptional Development and Freedom.(94)

capable of producing really good crops, but of withstanding severe and prolonged drought with surprising impunity. The fact is chalk absorbs an enormous quantity of water, and gives it back slowly and steadily when the soil above it dries. Furthermore, when the surface of the chalk is broken and some of the gravel and earthy matter is dug or washed by rain among the chalk the roots of a vast number of plants will readily delve deep into the cool moist area and with such access to moisture and some nourishment the plants can defy the hottest and driest summer weather, whilst the fact that in winter they will never suffer from stagnation and sourness keeps many plants in fine condition that succumb to winter damp in other soils.

There is less advantage in introducing clay to a gravel soil which is already pasty and adhesive after every shower of rain than in the case of a sandy soil which is loose and sharp. Humus is of course necessary, but it is frequently the case that despite the stone the gravel is sufficiently retentive to justify early application of animal manure, and for this purpose cow manure or farmyard manure is specially suitable.

Where chalk is near the surface deep digging usually ensures the maintenance of a sufficiency in the top spit, but at the same time it may be remarked that it is quite possible for a gravel soil to rest upon a vast mass of solid chalk and yet be seriously deficient. Lime in all its forms tends to work its way downward through soil, assisted in the process by every shower that falls. Therefore be not surprised though the eighteen inches or two feet of crust over a chalk base may be so lacking in lime that no shallow rooted plants attain a really healthy rich green tint of foliage.

Not very long ago my advice was sought regarding the health of a lawn growing on a shallow covering of earth over a bed of chalk known to be at least fifty feet deep. The condition of the grass convinced me that deficiency of lime was the root of the trouble, and the scepticism of the owner was such that it was agreed that only half the lawn should be limed. The result was so impressive that the second half was very soon dressed, and the owner went about advising everybody to lime their lawns. That is just the trouble; no sooner is lime seen to do good work where conditions justify its use than away goes the advice to all and sundry that lime should be used here, there and everywhere; and that advice may as easily prove to be mischievous as beneficial. Some grasses do not like lime and are best without it, and often an excess is much worse than a deficiency. In another garden resting on chalk a deep trench was made for the purpose of laying an electric cable. The chalk thrown out was distributed over a considerable breadth of the garden on either side, much to the annoyance of the occupant of the house. During the next season, however, he found that wherever the chalk had lain plants made more vigorous growth and leaves were of a darker green. The next autumn a deep hole was opened in a corner of the garden and sufficient chalk excavated to dress the whole garden, and the improvement in the ensuing crops was uniform throughout.

One trouble with very stony ground is that the stones afford shelter for many ground pests, especially slugs, woodlice and earwigs. Soil fumigants should be raked in the surface, or the soil may occasionally be drenched with alum water.

Where gravel rests on a hard, harsh bed of large flints or sandstone it is certainly advisable to haul out a quantity of the biggest, loosening the whole bed as deeply as possible and making up the cavities caused by the removal of the stone with rough material that will rot down to form humus. Where such rough stony beds are concerned one may tolerate the burying of fairly extensive masses of vegetable matter without using up good soil, but the buried material must only take the place of the removed stone and not increase the level to an extent that unduly raises the surface soil, for it is not good to lift a stony gravel much above the surrounding level, or the plants growing therein will be severely punished during a hot summer.

The benefits of green manuring must not be overlooked, and it is a noteworthy fact that leguminous plants will usually make good crops on gravel and chalk soils.

The great tendency of such soils to batter down during heavy rains and set with a cement-like crust must be counteracted by much hoeing. On large areas room should be left between straight rows of plants for a wheeled cultivator, and in smaller gardens the three or five-pronged "Buco" cultivator will be an implement of great utility.

CHAPTER IX

MARSH LANDS AND LOW-LYING BLACK SANDS

FREQUENTLY rich with deposits that have been left behind by countless flood tides, generally acid through a superabundance of humus, nearly always capable of growing remarkably fine crops if properly handled, marsh lands and black sands require expert treatment, and sometimes call for expenditure of considerable sums of money, which, however, they are capable of quickly recovering.

Drainage is usually of imperative importance, but the problem is how to dispose of the water, how to secure a fall and outlet when the land lies at or even below water-level.

Irrigation is in some districts carried out by means of channels or ditches at stated intervals throughout the land. Drainage may be similarly carried out, but the ditches must be deep, and must be designed and constructed by competent hands. The expense is heavy, and it would be folly to incur it without knowing that the correct method is adopted. One point I would urge: that the soil excavated from the ditches should be packed up in banks on either side the ditches rather than spread over the surface of the cultivated area. The surface soil may be rich and tolerably sweet from frequent disturbance and exposure to the air, but th.at which is brought up from the depths will be rank, sour, and unworkable for a long time. Quite possibly after a year or two of "weathering" the banks may become plantable with nuts, blackberries or something that will pay for the ground they occupy, and an exposed marsh divided into squares by a series of such ditches and banks will make an extremely productive well-sheltered holding for the cultivation of highly remunerative crops.

Land reclamation is a big and important subject that cannot be exhaustively dealt with in this book, but in various parts of Britain much land which was at some time reclaimed is now inhabited, and many home gardens rest upon such land. Their owners have the problem of proper treatment of their soil to consider, and it is well to emphasize the point that however much the soil may have improved, its fertility still depends upon maintenance of a good sound drainage system. The formation of a pond, sinking a well, deepening a ditch are sometimes means of averting trouble, and quite a deal of importance attaches to planting a certain number of trees, rushes, or other plants that will absorb a great deal of moisture, improving the condition of the soil around them to the advantage of other plants.

Lime is here again of great service; not, of course, as a disintegrator, for the black sands and marsh silts are usually loose and porous, but as a corrective of acidity and an antidote to an excess of humus which such soils usually contain. Caustic lime is most effective for this particular purpose, but ground limestone may sometimes be used to considerable advantage.

Cultivation of soils of this character would present no great difficulties if only they were immune from occasional floods or perpetual stagnation. The ditch draining is a preventive to a certain degree, but cannot always cope with phenomenal downpours or abnormal tides. Some landowners will aver that periodical floods are highly beneficial to land, but notwithstanding the fact that a great deal of manurial matter is deposited and left behind by the receding tides, that manure is often so rank and sour that a deal of harm is done to the surface soil. The saturation, too, is detrimental, especially where natural drainage is so sluggish that the land remains puddled for a lengthy period. Even these soils may not properly be worked while in a muddy state, for it is surprising how even the soft silts and black sands will knead into a dough-like paste and later become wind-dried and sunbaked into hard clinker-like masses. Often there is a complete absence of stone in these soils, and on various occasions I have advised carting clean shingle on to a plot of perfectly stoneless land, and where that advice (given after careful examination of the ground) has been acted upon it has proved to be more beneficial to the land than heavy dressings of expensive fertilizers.

Many times I have heard even commercial growers remark that their land will stand any quantity of manure and fertilizer every year, and that the only way to get anything out of it is to put plenty into it. Sometimes 50 per cent. or even 75 per cent. of the money thus spent might well be saved, and not only equal but superior crops be grown in consequence. Half of the errors in this direction are due to the dissemination of a certain amount of information regarding the scientific principles of land nourishment. The information may in itself be accurate, but its assimilation by the scientifically untrained mind is rarely complete, and certain isolated or disjointed facts are gathered which can only be of real service when collateral points are taken into account and carefully considered. Thus it frequently happens that the rich alluvial soils, highly impregnated with humus, are supplied with fertilizers whilst they already contain a superabundance of locked-up plant foods which only await liberation by application of lime. Sometimes, also, the greatest need is for a change of crop, and sometimes the alteration of texture by the introduction of shingle, limestone chippings, burnt oyster shell or bone charcoal would suffice to extend the period of fertility for years without further manurial replenishments. The mechanical improvement of soils of distinctive character is frequently of immensely greater value than the customary system of annual enrichment.

One plain fact confronts the would-be critic, whether he ventures to express views upon the defects and shortcomings of marsh lands and black sands or to condemn faulty methods of cultivation; that fact is the marshy lands are sufficiently productive to attract large numbers of commercial growers, and the growers in such districts as the fens of Lincoln and Cambridgeshires and again of the Wirral peninsula, and the Ormskirk district of Lancashire are among the most prosperous of horticulturists. There are probably more men in occupation of flourishing farms and market gardens in these marshland districts who started as labouring men without capital than in any other parts of the kingdom; and when men are successful it ill becomes strangers to assume the role of teachers who will quite easily turn their pupils from paths of error. Many such teachers have themselves been disillusioned, for the marsh land cultivators. have a knack of confronting theorists with very stubborn facts.

The prosperity of the marshes is due to a combination of the remarkable fertility of the soil and the untiring industry of the cultivators. There are, however, two points which even the most successful of this prosperous class of growers must acknowledge to be true; one is that their whole system of cultivation is very costly; and the second is that they are constantly troubled to a greater degree than other districts with seriously destructive pests and diseases; and in both these respects they are victims to their own stubborn adherence to customary practice.

On the face of it there is soundness in the contention that the method and system which have made them successful and prosperous must be correct; but deeper investigation and closer study will show that even good systems cannot be adhered to indefinitely without involving deterioration in results, and sometimes opening the way to unwelcome and serious troubles.

Because a particular manure or fertilizer produces good results for a season, two seasons or three, it by no means follows that it is good for the soil or the crops to persist in its use for twenty consecutive years. It may be added that change of brand may not necessarily mean change of character of the fertilizer; and that of course is the important point. When a plot is deficient in potash a dressing of sulphate or muriate of potash will be beneficial, but after a few years a change will be essential. When that change is made it will not serve the desired purpose to pass from sulphate of potash to Kainit, nor even to apply a guano that shows an exceptionally high percentage of potash.

The same remarks, of course, apply to phosphates and nitrogen, and therefore changing the fertilizer must be something more than merely the name.

There is no doubt much waste of money and injury to soil may be avoided by frequent alterations in feeding and also in cropping.

It is by no means uncommon criticism of the advocacy of rotation in cropping to bring forward instances of continuous cropping with potatoes, onions, or maybe daffodil bulbs, but such critics cannot prove that it would not have been more profitable to give the land an occasional change of crop, and they apparently overlook the fact that not only can soil be sickened and impoverished by the continuous occupation by one kind of plant or crop, but that when once a ground pest or disease gains a foothold its development is fostered and encouraged by continuing to provide it with its own particular food or host plant.

Whatever may be brought forward in opposition to the proposition that frequent change of both crops and nourishing agents of the soil is good cannot shatter the argument that such change is more calculated to be beneficial than detrimental; common sense must range itself on the side of such reasoning, and it is supported by the fact that we have seen diseases and pests become greater scourges where land is constantly nourished and cropped in the same way than where frequent changes in both are made.

CHAPTER X

MOORLAND AND SANDY PEATS

WHERE gardens have to be made on moorland peats and the loose, dry, sandy peats which cover large areas in certain parts of Britain the most satisfactory course is to adopt a style of gardening which enables selection of trees, shrubs and plants to be confined to such as make themselves contentedly at home in peaty soils. Even grasses for lawns may and should be specially blended for sowing peat-lands, for ordinary mixtures or ordinary turf will fare badly on soil which is entirely at variance with their natural requirements.

Familiar examples of suitable subjects for peat lands are rhododendrons, azaleas, kalmias, the hardy heaths, gorse, ferns, and a fairly wide range of ornamental trees and flowering shrubs. Rock gardens and herbaceous borders may be made and furnished with comprehensive collections of peat-loving plants, but the subjects which are unlikely to satisfy include roses, carnations, phloxes, and such plants as are usually described as requiring a good loamy soil. My habitual custom is always to advise that it is better to choose one's plants to suit soil and environment than to embark upon the expensive and sometimes extremely difficult task of converting an unsuitable spot into a comfortable home for plants of exacting requirements.

Peat lands are of totally distinct character from all other soils, and there are two distinct kinds of peat lands: bog-peats, which are dense, close, always saturated and suitable only for bog gardening, since they would not even make good garden soil if most perfectly drained; and the moorland or heath peats, which are seldom saturated for any length of time, because whilst the fibrous peat absorbs water like a sponge the presence in abundance of a powdery white or yellow sand allows water to filter through as through a Sieve.

Such peats are surcharged with humus, but are practically destitute of lime, and sadly lacking in the important elements of plant foods which are found in loamy soils. Hence the recommendation to make the best of things by planting peat-loving subjects, among which are many of our most glorious garden plants. Often attempts are made to specially prepare a patch for roses or some special favourites among loam-loving plants, but even where expensive excavations are made and refilled with imported soil the results are frequently akin to the efforts of dear old ladies to grow geraniums and fuchsias in cracked pudding basins. Despite these firm convictions, fully realize that there are enthusiastic amateur gardeners who at all costs are determined to grow their favourite flowers in the only gardens they are able to possess, and many gardeners in private establishments are expected to grow as good and varied a range of plants, fruits and vegetable crops on their peatlands as are grown in the most fertile loamy soils. For the benefit of such it behoves me to include in this book the best advice I may be able to give regarding methods of improvement of peaty lands.

Situation of such lands usually means that anything in the way of good loam or other requirements has to be carted a considerable distance, and it is a pity that labour and expense should be thus incurred and the benefits reduced to a minimum through mistakes in the method of procedure; but such mistakes have been observed on various occasions. Loam, clay, marl, or even friable garden soil, will not mix effectively with brick-like chunks of fibrous peat; and even if the chunks are broken or cut in pieces with. much labour the amalgamation is unlikely to make a good comfortable bed for fine-rooted young plants.

The alternative plan of filling holes entirely with imported soil cannot be recommended; it would be better even to plunge plants in pots rather than plant them in patches of soil that will become mud pies after every heavy rain.

The first steps in a reasonable method of converting sandy peat into a passably good garden soil have already been detailed in a previous chapter. Where imported clay, loam or marl is available it should be evenly spread over the ground between the rough ridges of peat, and left with them, exposed to a winter's weather before the peat is shaken out for burning. Then the ashes and the imported soil should be evenly spread, and the land well dug, taking care not to bring up more rough cakes of peat from below. As digging proceeds quicklime should be distributed along every spit, at the rate of about two pounds to every square yard.

Manuring should be postponed to a later digging, and planting should be delayed until spring is fairly well advanced.

It must not be anticipated that one season's treatment of a hitherto uncultivated peat will suffice to bring it to the condition of good alluvial or loamy garden soil which is capable of growing anything and everything. It may certainly be rendered fit to grow a serviceable crop of many kinds of vegetables. Potatoes and green crops will flourish if assisted with some quick acting fertilizers; and the potash provided by burning the peat fibre will be of great help to potatoes. So far as flowers are concerned, annuals and summer bedding plants will be the best subjects for the first season, more permanent planting being taken in hand the following autumn.

Such is the effective method of improving peat lands. The fibre of the peat in its original form is practically valueless to the average run of plants, but the ashes after burning will be rich and beneficial, whilst the removal of the fibre facilitates the proper incorporation of the clay or loam with the sandy solids of the native soil. Subsequently the soil will benefit by dressings of animal manure, and also of bone-meal, hoof and horn shavings, fish-meal or dried blood manure.

Top-dressings of bone-meal, fish-meal, Peruvian guano, etc., will benefit lawns on such a base, and in fact with a thoroughly good start on these lines the garden may be treated in the ordinary manner of good cultivation. In all probability light dressings of lime will be required about every third year, but intelligent watchfulness will be one's guide as to this matter. Loss of healthy green tints in foliage is a sign that the lime supply is exhausted and needing replenishment, and this will naturally be more frequent in the case of a very open, porous soil that was originally destitute of lime. The smallest need for several seasons will be humus, and for that reason sheep manure is preferable to stable manure or farmyard manure, whilst the ashes of burnt refuse are more valuable than decayed vegetable matter.

Two or three years of judicious surface treatment will make a remarkable difference to the top spit, but by that time another dressing of burnt fibre ashes will be of great benefit, and at the same time the subsoil will be all the better for another turn over. A capital plan will be to trench the ground, throwing out during the process as much of the coarse rough fibre and rubbish as possible, allowing its place to be taken by some of the improved surface soil. The rough material should be thrown in small heaps, and with the aid of gorse, heather, bracken, etc., which can usually be collected on moorlands, a series of fires may be formed to char the subsoil which has been thrown out.

Some peats are deep and thick, but some are only thin coatings over a bed of soft, fine sand, and some cover a harsh brashy conglomeration of stone and acid soil of a most disagreeable character. The last named must be deeply cultivated and limed, but must not be brought to the surface, for no benefit accrues from endeavouring to burn stone and grit. There will be all the more need here to bring in brushwood, gorse, rough grass and anything else that can be burnt to make good ashes, and with the help of sheep manure gradual improvement will be secured.

CHAPTER XI

OLD SOUR GARDEN SOILS

THUS far we have considered soils which require conversion from ill-balanced and more or less imperfect soils into fertile garden soils, but a very large number of gardens exists which alter prolonged periods of cultivation and regular manuring have passed from a state of rich fertility to a condition of rank acidity or sourness.

Such gardens are more common than might be supposed, for despite the general scarcity of stable and other animal manures there are cases where the garden owner has easy access to regular supplies and has year after year taken full advantage of his good fortune.

Frequently, too, in town gardens the most convenient method of getting rid of vegetable refuse, garden and kitchen rubbish and garbage of a miscellaneous character is to bury it in the garden, and this is cheerfully done in the knowledge that "It all makes manure." Quite so; it all makes humus, and the greatest fault of the majority of gardens that have been long in cultivation is that they contain far too much of the humus of which most new gardens are sadly in need. Whenever the humus in a garden soil exceeds 10 per cent. a condition of acidity begins to afflict it, and that condition is aggravated by every addition to the volume of humus. When a soil always turns up black, becoming greasy and pasty when wet, and losing the wholesome fresh smell which should rise while digging proceeds, it may be taken for granted that the humus is in excess of requirements.

To manure such soil is worse than useless; it is distinctly injurious. The great needs are rough digging and exposure to frosts; application of caustic lime, sometimes a liberal dressing of sea sand or other sharp sand, this necessity being conditional upon the degree of porosity of the soil; and if possible the burning of some of the subsoil to provide an inch thick coating over the whole area before digging over in spring.

Vegetable quarters should be for a season hard cropped with a succession of quick growing, gross feeding green crops and potatoes, and if convenient annual sunflowers, Shirley poppies, dahlias, hollyhocks and other plants of coarse, rapid growth should be freely planted in the flower-beds and borders.

Sour soil should not be allowed to lie idle, but should be taxed to produce rapid luxuriant growth, which, however, should be removable in a comparatively short space of time, in order that the soil may be again roughly dug and exposed to wind, sun, and again to winter's frosts.

Refrain absolutely from working or treading the soil while very wet, for nothing aggravates sourness more than churning up pasty soil or compressing it by treading.

THE DERELICT OR NEGLECTED TOWN GARDEN

Unfortunately all householders are not horticulturists, and those of the industrial classes who are keen gardeners sometimes of necessity enter into occupation of a house to which is attached a small patch of ground which should be the garden, but which the previous tenants have reduced to a pitiful condition. Trodden hard as stone with the trampling of many feet, soddened with water, perhaps, or polluted with spilled tar, paint, or other equally poisonous waste, choked with chips and shavings of wood, bits of glass, tiles and slates, the earth of the small town garden that has suffered a period of completely indifferent neglect is about as disheartening a problem for an amateur gardener to face as can well be imagined.

True, the optimism and pluck of the average gardening enthusiast is equal to herculean tasks, and the more limited the knowledge the more cheerily will the prospect be contemplated and the task of putting the garden straight be taken in hand.

The "putting straight" is perhaps not really

difficult, if only one has a stout fork, a pickaxe and a back which ignores fatigue, but in the ease with which the untidy plot may be made spic and span, and in the eagerness to get it done, lurks real danger that from the point of view of practical horticulture the first steps at improvement may in reality aggravate the already serious defects in the soil.

The hardened, compressed surface soil will probably be buried beneath the surface, there to remain in its deplorable, lifeless condition. The impurities and poisonous substances that may have been strewn upon the ground will probably become also buried, and the soil which, having lain beneath the hard trodden surface, has become vile and sick will be brought to the surface where tender young roots, seeds, etc., will come in direct contact with its evils.

Haste is fatal when dealing with these derelict or neglected gardens, for here above all is required a thorough cleansing by exposure to the air, by liming, and by deep trenching. If possession be taken of a neglected garden in summer or autumn there should be no thought of commencing planting before spring, and if it is spring when the ground is taken over a whole summer's preparation is necessary to get ready for planting bulbs, roses, etc., in autumn.

First look well over the surface and clear off anything and everything that is not soil. Then fork over the surface but not too deeply, and having done that, picking out during the process whatever there may be of rubbish, leave until the loosened soil is nicely dry and then make preparations for thoroughly trenching the whole area. Have lime ready to work in as the trenching proceeds, and apart from digging into the subsoil whatever is deemed most serviceable to open it up and sweeten it, reserve manuring for the top spit at the next digging.

Remember that in autumn the rougher the top spit can be left the greater will be the sweetening influence of air and frost, but if the trenching is done in spring break down the clods and leave the surface as loose as possible, lest the rough lumps become hardened by wind and sun, and further progress is hampered in consequence. The great essentials are to open up the subsoil to improve drainage and to expose as much of the surface soil as possible to the sweetening influence of the air, at the same time applying the lime to accomplish its beneficent mission of correcting acidity. In due course, but not too soon, in fact not until the surface has broken down and become again fairly close, dig over the top spit again, this time manuring, if possible, with good straw-manure from a stable, but not with that from stables where shavings or sawdust are used for bedding.

In town gardens it is usually a difficult matter to make garden fires for burning either rubbish or soil, but when it is possible to procure either burnt earth or good wood ashes make the most of these by distributing evenly over the soil, covering but not burying them too deeply.

Cinder ashes are very often buried in larger quantities than is good for soil, especially when, as is usually the case, they are brought to the garden fresh from the fires. Stiff soils like London clay will be benefited by a moderate quantity of coal ashes, but they should lie exposed to the weather for a few months before being dug in, and are more suitable for opening up the subsoil than for the top spit, where, if used at all, they should be limited to a very light sprinkling. Many town gardens lie low and hollow and are badly drained, but the difficulty of finding an outlet, and the fact that gardens on either hand will always minimise the benefit accruing from any effort at pipe draining will generally reduce one's practicable measures to the one method of digging a deep hole at the lowest end of the garden and filling it up with bricks, stones or clinkers. When this is done do not make the mistake of spreading the soil from the hole on the garden, but either make a heap, eventually to be planted with some strong growing plant such as hypericum, ivy, or periwinkle, or, if possible, remove it from the garden. One season's treatment on these lines will work wonders in the improvement of the worst of soils, but with each succeeding year still further improvement will be effected by surface digging, hoeing to improve tilth and facilitate aeration, and by judicious use of manures, lime, sand or other recognized necessities.

Because some old gardens attached to town houses have become overcharged with humus and soured through the frequent burial of putrid garbage it is not to be assumed that humus must never be added to the soil of a small town garden. In many London suburbs the natural soil is a particularly stubborn clay, and in these circumstances the addition of humus is quite as essential as elsewhere, and all effort should be made to improve the texture of such soil until its darkened colour and friability give indication that the deficiency has been made good. Tea leaves, pea pods, bean pods, cabbage leaves, lawn mowings and any waste greenstuff may be buried, so long as there is no suspicion of the presence of disease. When, for the sake of tidiness, all the garden and household rubbish has to be thrown into a pit to rot, it is well to cover layer by layer with a little ordinary lime. Soapsuds and house slops may be tipped into the pit, but before the rotted rubbish is dug into the garden it should be spread for a short time over the ground to lie under the sweetening influence of the air. During digging it may again be beneficial to sprinkle a little caustic lime along each trench. The notes on green manuring will be of service to the owner of a garden on London clay, and also to those who live in seaside towns or on rocky or chalky hillsides.

CHAPTER XII

MANURES

CORRECTLY speaking, the word manure embraces all substances which, are used or may be used for the purpose of increasing or restoring fertility in soils. Common custom in agriculture and horticulture is to use the term "manures and fertilizers," the former being assumed to apply to animal manures and the latter to chemical substances and compounds and to concentrated manures, such as guano, dried blood, fish-meal and various other plant foods of complex character. "Artificial manure" is another commonly used term, but by no reasonable line of argument can that be held to be correct or apprdpriate.

There is little use, however, in endeavouring to oust from general use terms which have by long custom established themselves, and it will very well suffice for the practical purposes of this book to deal separately with manures which are of the nature of animal excrement and those which are of easily definable chemical constitution, whilst a third group of miscellaneous character may for convenience be dealt with under the heading of miscellaneous fertilizers. Expert chemists may prove by the scientific means at their disposal that the finest and most profitable results are obtainable by treating soil with prescribed quantities of certain substances which are capable of restoring a correct balance of necessary plant foods; they may also prove that any other method of applying nourishment to soil is wasteful and therefore extravagant, and we shall not attempt to contradict their assertions, because we know that the expert chemist speaks with a profound knowledge of his subject. We implicitly believe many things chemists and doctors tell us, but even of that which concerns our own lives and health we believe but ignore much, for the simple reason that we are quite unable to order our modes of life so precisely that we could carry out to the letter the whole of a learned doctor's regulations for a model life, nor indeed does the doctor imagine for a moment that one might do so. For the most part the best we can do is to avoid the wilful and flagrant indulgence in that which is obviously fraught with immediate and serious danger to health, and somewhat similarly the most the average cultivator of the soil can do is to avoid the blatant errors of either starving or poisoning the soil in which he wishes to grow plants or crops.

Much teaching and much literature has for years been devoted to the subject of the chemistry of soils. The use of chemical manures has been strongly and eloquently advocated, and with such good purpose that thousands of tons of chemicals are applied to cultivated lands in Britain every year, and in the most rural of our villages to-day one may hear the humblest rustics conversing about phosphates and nitrates, sulphates and potash salts. Far be it from me to disparage or scorn the modern growth of the science of chemistry in relation to the treatment of the soil. Chemical fertilizers are a great boon to agriculture and horticulture, and he would be an utter fool who would condemn them or belittle their importance and value. The cause for real regret lies in the two vital facts that all cultivators of the soil are not expert chemists and that the nett result of all the literature and lectures which aim at teaching plain men of the soil the science of chemistry has been to endow large numbers with just that little knowledge of a vast and profound subject which is in a very real sense a dangerous thing.

The danger makes itself conspicuously apparent when, as so often happens, the person who has acquired his little knowledge credits himself and is credited by others with the possession of a great store of learning, for it is then that he assumes the role of teacher, with the result that his willing pupils are misled by a confusion of fact, supposition and error which passes for scientific teaching. Whilst, therefore, it may be readily granted that plants and crops can be sustained and nourished by chemical foods and that soil may be brought to and maintained in fertile condition by the aid of chemicals, it must also be insisted that an essential condition is that the manipulator of those chemicals must be a fully qualified analytical chemist. I do not hesitate to go further, and say that thoroughly satisfactory results from the sole use of chemicals are seldom obtained for a succession of several seasons, even when the supervision of a qualified chemist is ensured, but on the other hand much land has been seriously injured by persistent use of chemicals and deprivation of animal manure.

Against this we have the incontrovertible fact that many gardeners and farmers maintain their soil in good condition and produce excellent, profitable crops year after year whilst relying upon animal manure with the sole addition of an occasional application of lime, or perhaps wood ashes and soot. Often enough the results would be substantially improved had certain chemicals been used in conjunction with the animal manure, and such combination is sometimes capable of effective economy in initial outlay in addition to increasing or improving crops. Often also animal manure is wasted and sometimes soil which is already rich is surfeited with an overdose of manure when something of a chemical nature would have been highly beneficial. Many mistakes are made in regard to the use of animal as well as chemical manures; were it not so, there would be small need to write about manure at all, but the point of practical importance is that to dress soils with animal manures is entirely in accordance with Nature's plan; and so wise is Nature that she usually contrives to let the earth reap the benefits derivable from animal manures, and frequently herself counteracts evils that might arise from the erratic methods of inexperienced husbandmen. Ignorance is undoubtedly less dangerous when handling animal manures than when dabbling in chemistry.

There is no occasion to interpret this to mean that the cultivator of the land should be indolently content to remain in ignorance, for ignorance is always extravagant, and whether horticulture is pursued for profit or purely for pleasure extravagance should not be tolerated. It behaves every individual who controls a plot of land, however small, to acquire as much knowledge as possible, that he may work his land economically to the utmost advantage; but I would rather see the novice in horticulture first learning all he can about animal manures and then advancing to a study of chemicals than that the reverse course should be taken. We shall find that there is much worth learning in regard to even the most common kinds of animal manure.

STABLE MANURE

Although increasingly difficult to obtain, owing to the substitution of motor vehicles for horses, the manure from stables is still the most generally available of serviceable animal manures, and is of great value. Farmyard manure, when well made and containing in fairly balanced proportion the manure from a variety of animals and poultry, is richer in plant food than horse manure alone, but horses are kept in many places besides farms; and especially in towns it is often possible to get stable manure when farmyard manure is unavailable. The manure from horses is rich in nitrogen and is a hot, dry manure by comparison with that from cows or pigs. It is therefore of greater value on cold heavy soils than on light, hot and dry sands.

The solids of this and of other manures contain less plant food than the urine, a large proportion of the former being merely humus-forming material. It is consequently of importance that the manure shall be mixed with some kind of bedding material which will absorb as much as possible of moisture. Good straw absorbs about three times its own weight of liquid. Sawdust will absorb more, but unfortunately sawdust is detrimental to soil, and also causes deterioration in the quality of the manure through violent heating. Peat moss litter is a great absorbent, and will hold seven or eight times its weight of liquid. Peat moss is suitable for soils which are lacking in humus, but all things considered, straw manure is best for most soils.

Frequently good manure is spoilt by bad storage. When thrown in a loose heap and exposed to weather stable manure will heat so violently that practically all its ammonia escapes, and ere long what little virtue is left in the manure becomes washed away by rain. It is of very great advantage if manure can be stored under cover of a roof, and the preservation of the value of the manure will quickly repay the cost of a corrugated-iron roof over the bin. When each day's manure is thrown into the bin it should be well trodden, as the firmer it lies the cooler it will keep, Although a slight shower of rain on a loose heap of manure will cause violent heat, a tightly packed mass will keep cool if thoroughly well soaked, for the more water there may be in a tight mass of straw the less room there will be for air, and exclusion of air prevents heating.

One sees with such frequency the recommendation that "old rotted stable manure" shall be used that it might reasonably be assumed that the older and more rotten the manure the richer it will be; but that is really not the case.

Much of the nitrogen in the manure when freshly made will have escaped in the form of ammonia during the process of decomposition, or rotting down. It is wise counsel that old rotted manure should be used when it is to be brought into immediate contact with the roots of young plants or when seed is to be sown on the surface without delay, for the fibrous material provides an easily permeated root-run, and the mildness of the nutritive qualities of the manure will be just suitable for the young roots, whereas freshly made manure would be altogether too rank and violent. Where, however, a plot of unoccupied land is to be dug, fresh manure, with long straw thoroughly saturated with urine, will add far more to the fertility of the soil, and in the case of a stiff cold clay the warmth of such manure will be of considerable benefit. Thus in autumn, winter or early spring it is wise to use manure in a comparatively fresh condition; not, however, before some slight fermentation has taken place; but for digging into soil between growing plants in beds or borders during spring or summer or for seedbeds, vegetable quarters, etc., the old rotted manure is safer, and is also more easily buried.

For top-dressing lawns very old manure is necessary, and it should be rubbed through a small-meshed sieve and mixed with a quantity of sifted leaf mould or fine loamy soil. The method as well as the time of applying manure to vacant land is of importance. One frequently sees good manure spread on the hard surface of undug land, and left for a considerable time. It is true rains will wash a certain amount of virtue from the manure into the soil, but it is also true that some of its strength will vanish into the atmosphere. Often again one sees digging proceeding and thick, solid masses of manure being laid in the trenches whilst the soil which covers the manure is itself left without its share of the enriching substance. It is an extravagant use of manure to bury it in masses, and the most economical and beneficial method of application is to distribute it in such a manner that every small piece of manure is brought into direct contact with soil; in fact, the more thoroughly the two can be incorporated the greater will be the benefit.

The principal elements of plant food which become exhausted by the continuous production of vegetation are phosphorus, potassium and nitrogen, or the compounds phosphates, potash and nitrates. Stable manure of good average quality contains all three in small quantity, but although the actual percentage is small, it must be remembered that stable manure may be safely applied in rather large quantities, and in the aggregate a very considerable amount of plant food is thus worked into the soil.

The important thing is to make sure that the manure is not ruined before use by being either overheated or impoverished by long exposure to heavy rains.

COW MANURE

It is quite commonly supposed that cow manure is richer in plant food than stable manure; but as a matter of fact it is not so, for it contains a great deal more water and considerably less phosphates, etc. The term cow manure covers bulls and bullocks, of course, and it is worthy of note that beasts fed on fattening foods such as oil-cakes, roots, corn and meal produce manure of far higher quality than those which are simply grazing on pastures. In all cases, whether it be horse, pigs, cattle or sheep, the quality of the food they eat has a marked effect upon the manure, and while young animals are growing or female animals are breeding, their manure is inferior to that of adult male animals. Cow manure is of greater benefit to light sandy or stony soils than to stiff retentive clay, because the manure itself is wet, plastic and cold. It will help to hold sandy soil together, hold moisture and tend to keep the temperature lower during hot dry spells. When cow manure can be obtained from stalls where the animals have been bedded with straw it is an advantage, but the byres on dairy farms are usually brick-floored and are flooded down with copious supplies of water. The result, of course, is that the urine is diluted; but even so the drainings from a cow byre are usually very powerful liquid manure.

SHEEP MANURE

The strength or richness of nourishment in sheep manure is greater than that of horses or cows. It holds much less of water, and is hot and dry by comparison with either of the other manures. Sheep manure is seldom obtainable with straw or bedding, the usual custom being to rake up sufficient from pastures to use for making liquid manure. For this purpose it is particularly valuable both for ground crops and for plants in pots. The small bulk and concentrated character of the manure somewhat discount its value for the purpose of digging in; it makes but a very small quantity of humus, and as its chemical properties are readily soluble it is not quite an economical proposition to substitute sheep manure for either that of horses, cows or pigs.

There is, however, a proprietary plant food on the market which demands attention, and although another chapter is devoted to proprietary manures, I prefer to deal with this one here, because no secret is made of the fact that its basis is sheep manure and the only accompaniment is wool. The name under which this product is distributed is

WELLSON'S PLANT FOOD

First thoughts may be that sheep manure and wool is scarcely appropriate material to describe as a proprietary article; but Wellson's Plant Food is not simply a rough and ready mixture of the two ingredients; it is treated with special care, and by an elaborate process is brought to a standard of efficiency which entitles it to rank with the best all-round fertilizers in commerce.

My own experience with Wellson's Plant Food extends over many years, and having used it under varying conditions, it has satisfied me that here is a plant food which may be strongly recommended with confidence that it will justify the recommendation. For no other reason would I place my signature under these remarks or a similar eulogy of any commercial speciality.

In the first place the moisture, which is unnecessary weight and a likely cause of deterioration in the quality or strength of the fertilizer, is completely dried out; the manure is then ground to a fine powder by powerful machinery, and a suitable proportion of short combings of fleece wool is ground to fine particles and thoroughly mixed with the excrement. The result is a dry meal-like substance which can be distributed thinly and evenly over the surface of the ground with the greatest ease, and can as easily be hoed in. The fine powdery state and dry condition facilitate rapid absorption of moisture and the ready solution of the plant food the manure contains, thus making a fertilizer of prompt action. As the nourishment assumes a liquid form, however, it comes in contact with the tiny particles of wool, which soak up and hold a large proportion of the liquid, holding it as a sponge until the roots of plants may require it. In course of time the wool decomposes and yields a further supply of serviceable plant food, considerably extending the effective period of the fertilizer. It will be realized that a finely ground meal, every particle of which is charged with nourishment, requires to be used in only small doses to do work equal to many more bulky manures which contain much water and insoluble waste matter, and whether distributed under potato sets, dug into vegetable plots, used as a top-dressing for flowering or fruiting plants or bushes, or used as an enricher of potting soil, the material is extremely

Same Group of Economic Plants.(130b)

A Group of Economic Plants.(130a)

economical, providing the maximum benefits of sheep manure in a much smaller than normal bulk. In the photographic illustrations the contrast between plants unfed and fed with Wellson's Plant Food, other conditions being precisely equal, afford ample evidence that this carefully treated natural manure is remarkably effective, and a further strong point in its favour is that it is far safer in inexperienced hands than unfamiliar chemicals or mixtures of uncertain character.

PIG MANURE

Analysis of pig manure shows it to be rich in plant food. It holds a great amount of moisture, and is on that account most suitable for light, dry soils.

The urine of pigs is of weaker rather than stronger constitution than the solids, but it must be added that analysis of pig manures reveals wide differences in various samples, due partly to the degree of saturation, and largely to the age and feeding of the animals.

Allegations are frequently made that pig manure causes brown scab on potatoes, club-root on cabbages, smut on onions and various fungous diseases on flowering plants. There cannot be truth in these allegations, but it is quite reasonable to contend that where land is continually and heavily manured with pig manure, and especially when the manure is, used in a fresh condition, any of these diseases which may be present will be aggravated and will increase in severity. It is well that pig manure should be stacked in a heap and thatched with a thick roofing of soil after the manner of a potato clamp, and shall be held thus in store for a period of several months before being dug in the ground. Furthermore, it is prudent to change this or, in fact, any other manure periodically rather than persist in the use of the same manure for many consecutive seasons. A full-sized cartload of pig manure should make a sufficient dressing for five square rod of land.

When pig manure has been used two or three times on the same piece of land a dressing of lime will be beneficial, but this should not be applied simultaneously with a further dressing of manure.

THE MANURE OF FOWLS, GEESE, DUCKS, TURKEYS AND PIGEONS

All these manures are highly nutritious; so powerful, indeed, that they require handling with considerable care. To dig them into soil in the ordinary way is really an extravagance, and the best method of procedure is to store the manure in a perfectly dry shed, beating it down close and firm with a spade whenever additions are made to the store. After several months of storage the mass may be chopped down, pulverized, and put through a sieve, and should then be mixed with two or three timents bulk of fine dry soil or sand. Such a mixture will form a top-dressing fertilizer equal in strength to a good guano, and should be as sparingly used. From two to four ounces per square yard of the mixture of manure and soil or sand applied to the surface soil between growing plants at intervals of a month during the growing season will constitute a powerful stimulant to almost any class of plants. The maximum amount should only be given when the manure comprises no more than 25 per cent. of the bulk; or if stronger one application should suffice for several months for any but the most vigorous and free growing of plants.

FARMYARD MANURE

The term farmyard manure is generally accepted to refer to a mixture of all or several of the animal manures which have been passed under review. It is of very variable character, since there is no recognised or, indeed, practicable system of regulating the character or proportion of the mixture, which generally depends entirely upon the kinds and relative numbers of animals kept on the farm or establishment. Farmyard manure is generally and should be invariably very good, and considerably more beneficial to land than anyone of its individual ingredients. It should be suitable for practically all classes of plants or trees, and may be used with advantage on well nigh all soils except old garden soils which have become acid through excess of humus.

The greatest pity is that farmyard manure is far too frequently spoilt by sheer carelessness.

Very often when visiting farms where good stock is plentiful one may see the manure simply thrown into a hollow in the yard, left-loose to heat violently, exposed to all rain, which dissolves the soluble plant foods, the practically black, strong liquor running from the bottom to find its escape along cart-wheel ruts or to soak away into the surrounding soil. Often the cow manure, the horse manure or the pig manure is simply dumped en masse on one end or another of the heap, so that in reality there is no mixture of manures, but rather several different masses leaning together, but each still apart from the rest of the several kinds, the whole spectacle affording unmistakable evidence of the fact that the real importance and value of manure is unappreciated, and that it is just looked upon as the muck-heap and nothing more.

If the wasted wealth of neglected manure heaps on British farms could be accurately estimated the figure would be astounding and could not fail to rouse the whole agricultural community to prompt action, in order to check an appalling and totally unjustifiable waste, but such are our ideas if the fitness of things that whilst lecturers travel the country and literature is published and distributed in endeavours to teach the population of rural Britain the science of agricultural chemistry we have not yet brought home to farmers a realization of the gold they are losing through utter carelessness in regard to the manure which is made at great cost on their own premises.

No landowner who erects a farmstead should consider his task completed until he has constructed a proper manure pit with walled sides, concrete floor and covering roof; the manure made on the premises should be the recognised means of improving the value of the farm. No tenant farmer should rest content until provision is made for careful preservation of all the virtue of every bit of manure made, and if such means were invariably provided, hundreds of farms which to-day are impoverished would be fertile and well nourished, and it would be no uncommon occurrence for farmers, after their own requirements have been adequately and fully satisfied, to have a substantial surplus of manure for sale, which garden owners-commercial and private-would eagerly purchase at substantial prices.

The trouble in the country is not that manure is not made, but that manure is so flagrantly wasted and that instead of securing whole manure a large proportion of that which is obtainable is but little more than empty husks.

It should not be considered waste of time to put men on to the task of turning and mixing the manure in the pit, but should be recognised as being just as essential as ploughing and harrowing the fields. When additions are made to the manure heap it should not be a matter of indifference where the cart or barrow is tipped and how the manure is left, but distribution should be considerately regulated, that blending of the various kinds may be facilitated, and the fresh manure should be trodden and soaked to check heating. The liquor should be carefully preserved with due appreciation of its intrinsic value, and ere long it would be found that twenty loads per acre would be more effective than forty had been under the old policy of careless neglect.

There is little risk of going wrong in regard to a dressing of well blended and preserved farmyard manure, whether soil be light or heavy; and so far as cultures are concerned, it matters not whether fruits, flowers or vegetables are planted, the nourishment of farmyard manure is bound to be acceptable.

However extensively and minutely one may enter into detail in written instruction, it will still be incumbent upon the gardener to use discretion, and to endeavour to regulate his own plan of campaign according to individual circumstances. Newly-broken ground which has not previously been a garden will naturally benefit by heavier dressings of manure than land which has been long cultivated and liberally treated, and the important question is not so much concerned with the quantity of manure that may with safety be buried in the soil as with the method of procedure by which the maximum of benefit and minimum of waste may be secured. For example, one will not bury good farmyard manure in rank harsh subsoil if small surface rooting plants alone are to be grown; manure will not be dug in early autumn into a light sandy soil which is porous as a filter if intending to postpone planting until spring. Where bulbs which like a rich soil but abhor fresh manure are to be grown the manure will be dug in six months before the planting season, and a green-manure crop taken to conserve the nourishment whilst dissipating the rankness of the manure. Due thought will be given to the possible need of lime, but great care must always be taken to bring no lime in contact with good manure while it is exposed to the atmosphere.

Such are points which, with the help of the teaching of this book, every individual must determine according to local conditions, but the vital truths I wish to drive home are:

There are manures of little as well as great value.

Therefore it is not all-sufficient to simply obtain and dig in whatever comes to hand as manure.

However good manure may be, and however cleverly it may be used, it can only accomplish a certain amount of work, and is alone insufficient to ensure complete success, which is the reward only of all-round good treatment of the soil.

CHAPTER XIII

SOME USEFUL MANURES OF VARIED CHARACTER

CLASSIFICATION of manures may be made the subject of much wrangling; but whilst argument may be very interesting, it may frequently be less profitable than a discussion on the utility of the materials within reach. My title is decidedly unclassical, but it will serve practical purposes, and is used in preference to the frequently used term "organic manures," because in point, of fact the great majority of manures or fertilizers are organic, even though we sometimes see lists divided into chemical sections and organic sections. Quite a large number of substances other than dung of animals are serviceable manures, and some are as useful in their capability of manufacturing humus as the manures with which the previous chapter was concerned.

SPENT HOPS

Brewers' spent hops are of great service in horticulture. There is not a high percentage of plant food in them, but they make a great deal of humus, and particularly on cold, heavy clay soils the hops will greatly improve texture and assist in improving both temperature and tilth.

Again, on light excessively porous sands the hops supply sadly needed humus, and their spongy, absorbent nature makes them valuable as a retainer of moisture.

It is of greater advantage to very light soils to supplement a dressing of hops by a good dressing of pulverized clay or marl, but greatest of all the advantages of spent hops is that they encourage extensive root development and will sustain quick, soft growing plants such as pansies, begonias and a wide range of flowering plants which would otherwise suffer and frequently collapse during severe drought. In the vegetable garden spent hops are specially suitable for lettuces, spinach, parsley, beans, celery, etc., and they are grand mulching material for raspberries.

A few years back it was my good fortune to possess a garden near at hand to a brewery from whence hops were readily obtained, whilst at a neighbouring dairy farm a tank into which the urine and wash from cow-byres and pig-styes drained provided an abundance of strong liquid which was placed at my disposal. A large flat-topped stack of spent hops was kept under an open fronted cart shed, and upon this at frequent intervals the liquid from the cowshed tank was poured and allowed to saturate the hops. This provided about the best manure it has ever been my good fortune to handle in ample quantities, and if by any chance the reader has opportunity to stack spent hops and saturate them with urine or liquid from a manure heap may he be advised to take full advantage of the opportunity and rejoice in his privilege.

Various commercial firms, among them being the well-known names of Wakeley Bros., and Abol Ltd., treat brewers' spent hops with a judicious mixture of chemicals, which the hops absorb, thus making what may reasonably be called a complete manure. Owners of small town gardens inconveniently situated for obtaining stable or farmyard manure may use these treated hop manures with every confidence, and there is a decided advantage in the fact that such preparations are free from the objectionable odour of ordinary manure.

There is just one point about hops; they are liable to foster various ground insects, including slugs, and it is advisable to dress the soil where hops are freely used with some such soil fumigant as Gasonite or Vaporite, or to sprinkle the surface occasionally with either soot or powdered lime, but never the two simultaneously.

FISH MEAL

A powerful fertilizer is made from decayed fish and fish offal. It is very rich, and may be used, but sparingly, on soils which are hungry for plant food but not lacking in humus. The presence of of a fair quantity of humus is necessary, because otherwise much of the virtue of the fish meal will be dissolved and washed away by rains before plant roots can avail themselves of the nourishment.

Certain rough grades of fish manure themselves contain a considerable amount of humus, but these are of lower feeding value than the finely granulated meals. Easy accessibility sometimes leads to a more frequent and liberal use of fish manures upon land near the seashore than is prudent. Salt and phosphorus, of which the fish manures possess a considerable percentage, are useful plant foods in moderate quantity, but in excess they become injurious. Seaside gardens that have been heavily manured with fish will be greatly benefited by green manuring, either mustard or vetches being used for the purpose, and it may be necessary to use lime as a corrective, but if the soil is very sandy and is not well supplied with humus the lime should not be caustic.

Inland gardens will frequently show great benefit from a dressing of fish manure, and also from pounded or burnt sea shells, oyster, cockle and mussel shells.

SEAWEED

There is a good deal of manurial value in seaweed, which contains a large proportion of salt and potash, together with smaller quantities of nitrogen and phosphorus. The tissue of seaweed is gelatinous, and rots quickly when buried or withers quickly when exposed to wind and sun; it does not, however, make humus to any appreciable extent.

The chemical constituents of seaweed are rapidly soluble, and being most frequently used on land near the sea, where the porosity of the soil ensures quick drainage, the benefits of seaweed are quickly realized, but very soon exhausted. Hence the frequency with which it may be used without detriment to the land. There is no doubt that much of our inland soil would benefit very considerably from an occasional application of seaweed, and on the more substantial soils the length of time during which its nutriment will remain available to plants will be greater than on the loose, light seaside sands. When filled with moisture seaweed is too bulky for cheap transport, but it may be reduced by drying to about one third its normal bulk without detriment to its feeding value. In such a condition, and in view of the fact that it may be obtained and dried with but little expense, seaweed would be worth carting considerable distances for the purpose of enriching the soils of inland gardens. Generally speaking, however, it is more suitable and beneficial to soils of comparatively light, porous texture than for stiff retentive soils. The latter generally contain potash, which only needs liberation by being brought in contact with lime, and so far as salt is concerned its tendency is to increase plasticity and to hold up moisture, which, of course, is quite an undesirable tendency in anything to be used on a cold wet clay.

HOOF PARINGS AND HORN SHAVINGS

Both hoof and horn possess manurial value of a high order, and whether obtained in the form of parings from the blacksmith or the finer waste from factories-separately or in combination-hoof parings and horn shavings may be used with great benefit to either heavy or light soil. The substance is very tough, and the process of decomposition is slow, and in that fact lies special value for certain purposes which may be readily explained.

When lawns are made it is essential that the soil beneath there turf or seeds shall be enriched, and whilst it is undesirable that a bulky manure which will rot and consequently cause considerable subsidence of the soil shall be used, it is essential that the manure used shall be of lasting benefit, for the simple reason that when once the lawn is laid the opportunity of gaining access to the soil beneath is past.

Hoof and horn manure meets the requirements of the case in an exceptional manner. Likewise when vine borders are made, wall fruits or other fruit trees and bushes are planted, rose beds, pœony beds, and in fact when any features of a permanent character are in course of construction hoof and horn manure may be dug into the soil with a knowledge that it will provide nourishment for a longer period than most manures.

It should not be too liberally used, six to eight ounces to the square yard being ample for even strong rooting trees, etc. For mixing with potting soil for chrysanthemums and other plants of substantial growth hoof and horn is a valuable fertilizer, four ounces to a bushel of compost being a sufficient quantity to use for the final potting.

BLOOD

Blood from slaughter- houses contains a good percentage of plant food. It is used by some growers as a liquid manure, but requires considerable dilution, and is abominably malodorous. Indeed, it is unfit for use in gardens surrounded by dwelling-houses, and as dried blood fertilizer is obtainable as an article of commerce and is far less offensive, but equally as beneficial to plants, there is no sound reason why townsmen should make their gardens a source of annoyance and discomfort to neighbours by using so objectionable a substance as crude slaughter - house blood.

Where land is situated far from human habitations a good plan is to prepare a bed of leaves, spent hops, loose soil or chaff, and saturate this with the blood as received from the slaughter-house. Cover the bed with a layer of soil several inches thick, into which a good quantity of gypsum has been worked, and after a period of three months or thereabout the whole may be wheeled on to the land and dug in. This will prove to be a very nutritious fertilizer for all strong growing plants or crops.

Dried blood fertilizer may with great advantage be used as a top dressing for either flowering plants or vegetable crops. It has considerable effect upon colours as well as substance of flowers, and will brighten and intensify reds, pinks and purples, but will sometimes cause whites and delicate shades to lose their purity. Growers of large onions and leeks for exhibition and also dahlia growers find dried blood a very serviceable fertilizer which promotes and hastens development of large dimensions.

SEWAGE

The sanitary authorities of many towns have a system by which sewage is dried, pulverized and made into a marketable fertilizer of rich character. It is, however, suitable only for farm crops, orchards, vegetable plots or allotments, and cannot well be recommended for gardens adjacent to houses. Dug or ploughed into the land in autumn or early spring it will produce good crops of corn, roots, potatoes, etc.

SUNDRY WASTE PRODUCTS

Waste materials of a varied character, such as leather dust, shoddy, hair, feathers, cloth waste, wool combings, etc., are sometimes recommended as fertilizers. They certainly contain plant foods in varying degree, but there is considerable irregularity in their worth and in their solubility. That a use should be made of everything is a good maxim, and one does not wish to injure the prospects of any commodity that may be turned to account. It would, nevertheless, be difficult to urge the advisability of placing reliance upon such miscellaneous waste materials for ordinary garden purposes, and whilst there is no lack of real manures and highly efficient fertilizers on the market, there is small need for the home garden to be made the burying ground for the refuse of all kinds of factories.

BONES, BONE MEAL AND BONE CHARCOAL

Here we have important and valuable material for enrichment of the soil and nourishment of plants. Bone is chiefly valuable on account of its phosphates, a good sample containing from 45 to 50 per cent. phosphates and about 5 per cent. nitrogen. Decomposition is extremely slow, and whilst this is an advantage for vine borders, orchards and such places as cannot be dug, it is a disadvantage for ordinary flower garden and vegetable garden purposes. Commercial brands of bone meal are treated in various ways with a view to hastening decomposition and liberation of plant foods, and there are wide differences in the relative values of various grades of bone meal.

It should be noted that in their original condition the phosphates contained in bones are insoluble, and it is only during the process of decay that these phosphates are rendered soluble and converted into available plant food. Naturally small particles buried in soil decay more rapidly than big pieces, and as a logical consequence the finer meals and small quarter of an inch pieces are more suitable for plants which require quick nourishment, whilst for vine borders, fruit trees, and other permanent features half inch and inch cubes are appropriate. It will also be apparent that bone meals are not of great value for top dressing planted beds during. the growing season, nor are bone manures of use for making liquid manure.

Green or raw bone is richer than cooked bone, but their properties are still slower in regard to solubility. Clay soils will benefit less than the lighter soils, and peat is particularly amenable to improvement by application of bone meal. The quantity to use on an average plot of land may be about seven pounds to a square rod, and when making a vine border that quantity may be allowed for two cubic yards.

Steamed bone meal differs considerably in analysis from that which is broken down by acids, but the latter, frequently termed vitriolised bone meal, gains in point of rapid decomposition. Bone ash, a white powder obtained by burning bones, is a powerful fertilizer, being very rich in phosphates. It makes an excellent top dressing for all kinds of plants and trees at the rate of four ounces to the square yard. Large quantities of bone ash are imported from South America, but the greater part of it goes into the factories where proprietary fertilizers are made up. It would be of great advantage to bring bone ash into more general use as in independent fertilizer.

Bone charcoal is produced by different methods, the bones in this case being heated in retorts. The actual purpose of its manufacture is to meet the requirements of sugar refiners, but it is a fertilizer of exceptional value, and a corrective of acidity in old garden soils and in moorland peats. Phosphates in some samples form as much as 70 per cent. of the whole, and it is the more valuable because freed of the organic matter of the bones. Seven pounds to the rod makes a serviceable dressing.

WOOD ASH

There is great value in the ashes of burnt wood, and of vegetable growth of all descriptions, the chief element being potash, of which a clean sample may contain anything up to 15 per cent. the residue being largely composed of lime.

For light sandy, peaty, or gravelly soils wood ashes may be used with a free hand, but there is less advantage in using this material on cold clay soils, for these are not usually in need of potash, and the tendency of wood ashes is to increase plasticity of the soil. Much wood ash is practically ruined by allowing it to lie exposed to rain, which dissolves and washes away the valuable potash salts. If the ashes cannot be stored in a dry shed they should be thickly covered with soil, forming a steep-sided clamp.

GUANO

Three quarters of a century or rather more ago guano was first imported into this country, and for awhile cultivators went into raptures over its magic powers, and in less than a dozen years from the arrival of the first shipload a season's importation reached almost a quarter-million tons.

Real guano is composed of the excrement of sea birds combined with decomposed feathers, the rotted carcases of dead birds, broken eggs and their shells, and parts of fishes upon which the sea birds feed their young.

The rocky cliffs and coasts of Pacific and Mediterranean islands were at one time coated. with thick layers of guano, but the seemingly inexhaustible quantities soon showed signs of being unequal to demand, and after a time some over grasping merchants began adulterating the guano with vadous make weight materials of little or no value to soil or plants, and other manufacturers of concentratedm anures or fertilizers began to use the name guano for materials that had about the same relation to real guano as cyder has to champagne. The consequence was that the period of excited enthusiasm for guano was followed by one of suspicion and violent prejudice, and for awhile there was a slump in the guano trade which cost many importers and merchants heavy financial losses. Good sense prevailed, and for some years past it has been possible to purchase genuine guano, unadulterated and unspoilt, and by law one is entitled to know before purchasing what is the analysis of the particular grade supplied. This latter is of material importance, because the age, density and position of the deposits affect the actual plant food value of guano to an extraordinary degree.

Some guanos are termed phosphatic grades because the percentage of phosphates is high and that of nitrogen is low. When this order is reversed the sample is termed nitrogenous guano. Nitrogenous guanos are worth more than phosphatic because in the first place nitrogen is itself more valuable and in the second place a high percentage of phosphates does not necessarily mean much plant food because much of the phosphate is commonly in quite an insoluble form.

Some "manufactured" guanos are cleverly handled by experts who first analyse a raw guano and then correct inequalities, adding potash, ammonia, etc., as required or treating with acid in a manner that will convert insoluble phosphates into soluble form.

Guano of a good grade is about the best all

round fertilizer to use for general purposes. It is rapid in action, and since it has but small effect upon the physical or mechanical condition of the soil it may be used wherever the need is for speedy nourishment of growing crops. Obviously the most economical method of application is by top dressing the soil during the season of growth, always hoeing the guano into the surface soil. For lawns, and also for beds of annuals and of vegetables which are to mature where sown, it is permissible to distribute guano during the final digging of the top spit before sowing the seeds, but unless the soil contains sufficient of spongy matter or humus to retain a good deal of moisture there is risk of a considerable portion of the rapidly soluble plant food disappearing before the young plants are able to make use of it.

Two ounces of guano per square yard is usually a good dressing, but the' quantity should really be regulated in accordance with individual analysis. Those who prefer to apply nourishment in liquid form may use at the rate of half an ounce to a gallon of water.

CHAPTER XIV

CHEMICALS AND THEIR MANURIAL VALUE

GENERALLY speaking the amateur gardener may with advantage both from the point of view of safety and convenience confine his use of chemical manures to the various proprietary compound fertilizers which are not only blended in accurate proportion by qualified chemists but are ground and mixed by machinery to a degree of perfection which is difficult to obtain when dealing with small quantities by hand. Nevertheless there are certain chemicals which are available in convenient form, sometimes at quite cheap rates, and which are capable, if carefully used in conjunction with or subsequently to animal manures, of doing excellent service, and it is appropriate that these shall receive due notice in this volume.

Familiar terms in horticulture are potash, superphosphates, ammonia, but there are wide differences in the forms of these excellent manurial substances, and one requires to exercise discretion as to whether, for instance, potash shall be applied in the form of sulphate, muriate, nitrate or carbonate or whether mineral phosphates will justify use instead of superphosphate of lime.

Much care needs to be exercised in the mixing of two or more chemicals, for whilst it is frequently more profitable to apply certain combinations than to use one chemical only there is great possibility of destroying the power of one substance by bringing it in contact with another, and it by no means follows that because each of three or four chemicals is capable of supplying a different element of plant food a combination of all will be correspondingly beneficial.

AMMONIA

Sulphate of ammonia is one of the most popular fertilizers in commerce, and being rich in nitrogen its value is not to be despised. One chief reason for the great popularity of sulphate of ammonia is that being readily soluble it is quickly taken up by growing plants and has an almost immediate, visible effect upon both the size and colour of their leaves.

A starved crop of broccoli or kale, spinach or lettuce, and likewise any flowering plant which normally makes luxuriant foliage but is puny and of sickly colour by reason of hunger for nitrogen will respond with wonderful alacrity to a reasonable dressing of sulphate of ammonia, given as a light sprinkling in dry form or in liquid.

Naturally a fertilizer which within a few days converts stunted unhappy looking plants into fine vigorous and healthy looking specimens with brighter green foliage quickly wins favourable regard, but there is need here for just a word of caution. Much that is highly beneficial if used at the right time and with reasonable restraint may become injurious when used to excess, and sulphate of ammonia is frequently used to excess, and may in fact be stated to be an unsuitable fertilizer for quite a considerable proportion of garden plants. As already stated, it stimulates and forces a rapid growth of foliage, but the growth produced is invariably soft, and if the chemical is used extravagantly the staying power of that growth is undermined and more or less destroyed.

Sulphate of ammonia dissolves on or near the surface of the soil, but loses a deal of its virtue before it penetrates to any considerable depth. It is consequently of greater benefit on cold, heavy and wet lands because it tends to attract roots near the surface where they get the benefit of warmth and air. It is unsuitable for light, dry sandy soils because in their case surface rooting endangers the plants to severe punishment during drought.

For crops of the cabbage tribe, for salads, spinach, soft wooded flowering plants and rapid climbers, especially when growing on heavy soils, sulphate of ammonia is a very excellent stimulant, and may be used during the growing period at a rate of from two pounds to four pounds per square rod, or the plants may be well watered with a solution at the rate of half-an-ounce to a gallon of water.

It is not a good fertilizer for plants of hard wooded character nor for vegetables that are to be stored for winter, for whilst it induces great size the bulk is produced at the expense of solidity, and flabby roots do not store well.

Growers of ferns frequently use sulphate of ammonia, and in moderation this is good, for it improves both the size and colours of the fronds, but if used too lavishly a degree of softness is produced which brings about the early collapse of the ferns, especially if they are removed from the congenial atmosphere of a well controlled fern house.

A wise plan is to use small quantities of sulphate of ammonia as a supplement to a light dressing of animal manure, or alternatively to use it in combination with superphosphates, sulphate of potash, or nitrate of soda. It should never be mixed with either lime or basic slag. For grass lands it is frequently recommended that a top dressing of sulphate of ammonia be used, and there are circumstances in which the advice may well be followed. For stimulation of rapid growth on pastures it is excellent; for checking development of clover in lawns it is effective, and for encouraging growth of grass on moss infested lawns it is also good.

For tolerably healthy lawns, however, sulphate of ammonia is less suitable, because it promotes rankness in the coarser grasses. The chemical should always be mixed with four times its bulk of either soil or sand, because if used alone it will badly discolour the grass. Never more than two ounces per square yard should be used, and since its sole effect upon grass is that of a stimulant, it is wise to use it once only in spring, and to follow with an autumn dressing of a more substantial fertilizer, such as Peruvian guano or fine bone meal.

Other forms of ammonia possess capabilities which are worthy of note, but being comparatively costly they are, generally speaking, only to be recommended for special purposes where large quantities are not required.

Nitrate of ammonia is a highly concentrated form, and is a combination of nitric acid and ammonia. Its chief value is to produce a fine healthy green in foliage plants, and exhibitors will find it of considerable service.

Where plants become yellow or pale green through lack of nitrogen a solution of half-anounce in a gallon of water will work a wonderful change in the colour of the foliage. It is particularly to be recommended for use in small green houses or conservatories attached to dwelling houses, for it is clean and odourless. Twice in a month is as often as this powerful stimulant may be used with safety.

Carbonate of ammonia is another form which should be used only in the form of liquid manure, the strength also being limited to half-an-ounce to the gallon.

Phosphate of ammonia contains soluble phosphates as well as ammonia, and on that account it is a valuable food for fruits, especially those grown in pots.

Flowering plants may with advantage be once watered with a half-ounce to the gallon solution just as buds are arriving at the stage when they may be expected to show colour. It is frequently of advantage, when flowers are required for a certain date, to be able to hasten such as are slightly backward. Phosphate of ammonia is capable of making several days' difference in the opening of most flowers.

BASIC SLAG

Steelworks provide us with this commodity, which is a bye-product of the industry of converting iron into steel.

Phosphoric acid and lime in combination forms basic slag, and when ground into a fine powder it forms a powerful phosphatic fertilizer.

Particularly suited for use on wet, heavy land, or for the black sands of marshes, basic slag should be used in autumn on land that is to be cropped the following spring. It is wiser to use basic slag independently of other manures or fertilizers, and the quantity may vary from seven pounds to fourteen pounds per square rod, according to the estimated deficiency of phosphates in the soil.

Basic slag is frequently recommended as a suitable top dressing for lawns, and on heavy soils it is decidedly capable of promoting dense growth of herbage. There is, however, the drawback that it encourages rapid development of clover, to the benefit of the farmer's crops, but often to the chagrin of the owner of a tennis court. Altogether, basic slag is of greater benefit to the farmer and to the maker of a new garden than to the owner of an established and planted garden, although it may be added that where a considerable area of vegetable garden or nursery land which has been long under cultivation is to lie idle for the winter months a dressing of basic slag is economical and beneficial, always excepting light, dry soils.

COPROLITES

One of the mineral phosphates, coprolites abound in Suffolk, Cambridge, and various other places, and when ground to the finest possible powder provide about 30 per cent or slightly more of phosphate of lime. Hence coprolites are, when obtainable without great expense in transport, a useful dressing for agricultural and market garden land. For the home garden, however, a good grade of super phosphates is of greater service; coprolites being as a matter of fact largely used by fertilizer manufacturers as a base material.

On light soil eight to ten pounds of coprolites in conjunction with four pounds nitrate of soda and four pounds sulphate of potash will make a substantial fertilizer.

GYPSUM

This is a substance possessed of serviceable qualities, but is not in itself highly nutritious. Sulphate of lime is another name for gypsum, and its constitution is lime, sulphuric acid, and water. Its action is similar to that of ordinary lime or chalk, and usually where chalk is readily obtainable it is to be preferred as a dressing for light soils. Gypsum acts like lime in liberating potash, but perhaps its most valuable capability is that of fixing ammonia to such a degree that if a manure heap is thinly covered with gypsum it prevents evaporation of ammonia from the manure.

IRON SULPHATE

After some hesitation sulphate of iron is brought into brief consideration, for whilst its serviceability as a fertilizer is not great there are advocates of application of iron to soils who have led many to suppose that it is never amiss to use a little sulphate of iron and even that a "good dressing" is an occasional necessity.

Plants require iron, but with very few exceptions the need is for minute quantities, and most soils have ample if not a superabundance for these needs.

Chalk soils, however, are frequently destitute of iron, and if magnesia is also absent, as it very likely is, all plants, will grow in a stunted, half starved manner with unhealthy yellowish green foliage.

Nitrogen does not always, under such circumstances entirely, remedy the defect, and when this is found to be the case it may safely be decided to apply finely powdered sulphate of iron at the rate of an ounce to the square yard, working it just beneath the surface with a Dutch hoe.

On lawns where toadstools and similar fungi are troublesome, or where fairy rings damage the grass, a solution of one ounce to three gallons of water may be used to good effect. In the case of fairy rings the circumference of each ring should be well saturated, whilst for toadstools, etc., a light watering may be given the whole lawn.

KAINIT

By reason of its cheapness, Kainit, which is a potash salt imported from foreign mines, is serviceable for extensive areas of land which are deficient in potash.

It should be applied in autumn, for its conversion into available plant food is slow. On this account it may be used with safety on bulb lands at planting time, for it will not hurt dormant bulbs and will become available just about the time developing buds in spring require additional nourishment.

Most vegetable crops benefit by an application of Kainit, potatoes, green crops and root crops alike being capable of taking up the potash during the season of growth following an autumn dressing. For fruit trees, bush fruits, loganberries and raspberries Kainit is a good food, seven pounds or eight pounds per square rod being a suitable dressing.

In flower gardening much finer results, except perhaps in the case of bulbs, are obtainable by using other forms of potash, such as sulphate, muriate, and carbonate. Kainit should be used on light, fairly porous soils rather than on plastic and moisture holding lands.

NITRATE OF SODA

Here we have one of the most important nitrogenous fertilizers at command; one that is capable of greatly improving a wide range of garden occupants when used with judicious care, but which may do serious mischief in careless hands.

The soda is rapidly soluble, and its effect upon growing plants is prompt; leaf growth being forced in its pace to a remarkable degree. Nitrate of soda is essentially a stimulant to use during the period of rapid growth. It should never be given to perennial plants late in autumn, as it will induce soft growth which will be unripened by the period of severe frosts, and by retarding the natural process of sap recession will prevent many plants assuming the dormant state which is necessary to their well-being.

For the purpose of giving a fillip to plants that from varying causes may have become backward and poorly developed, nitrate of soda is particularly useful. It is excellent for promoting succulent growth in salads and soft green vegetables, in developing celery, seakale and cucumbers, and in improving the foliage of pot plants.

The best method of application is to sprinkle one ounce to the square yard over the soil around the plants immediately after a shower of rain. Two or three weeks later a similar dressing may be given, and that should suffice for most things. It is not advisable to apply nitrate of soda to plants in full bloom, as the result will be to soften the substance of the petals and hasten their collapse. A convenient method of applying this stimulant to pot plants is to dissolve one ounce in three gallons of water, and at that strength it may be given to fairly vigorous plants three times in a month.

To use nitrate of soda on land which is poor, or in other words deficient in potash and phosphates, is an error, because the action of the soda will be to increase the demand of the plants for the other elements, with the result that the soil is still further impoverished. A dressing of animal manure or of phosphates and potash should follow any considerable use of nitrate of soda.

POTASH.

So incessantly is it reiterated that healthy plant growth is impossible without potash that there would be little justification for harping upon that simple fact. There are many plants and crops grown which could very well make good use of more potash than is at their disposal, nevertheless it frequently happens that potash is bought and, applied to land which already contains ample supplies, but in a form which is useless to plant life. Clay soils, for instance, have usually enough potash to serve any plants that might be grown in them for several years, but being locked up in an insoluble form, the food is unavailable to hungry plants. Application of lime effects the desired change in the character of the existing potash, and when clay soils are limed it will generally be needless extravagance also to apply potash. Wood ash contains carbonate of potash, unless the ashes have been left exposed to rains until the virtue has been washed out, and where a good dressing of wood ash can be dug in or hoed into the surface of planted areas there should be no further need to purchase or apply potash for at least a year. Kainit contains agood per centage of potash, and some grades of guano are so rich in potash that additional supplies are not required. For general use sulphate of potash is the customary form, and there is no question of its utility and all round merit.

The commercial article varies a good deal in point of strength or purity, and when purchasing it is well to see that analysis shows something closely approaching 50 per cent. potash. Sometimes inferior grades are offered which contain only about 30 per cent., but vendors are under obligation to declare analysis, and the best grades are in the long run cheapest.

An advantage regarding potash is that although readily dissolved by rain, any soil which is fairly well supplied with humus is capable of holding up the plant food, and it is safe therefore to apply sulphate of potash at practically any season of the year.

Four pounds per square rod may be considered a good dressing, and there are few plants which may not be fed with potash, it being specially required by potatoes, tomatoes, and root crops.

Muriate of potash is equally powerful with sulphate, but it is less generally used. For sandy soils muriate of potash is more suitably used than sulphate, its action being somewhat steadier and more sustained and its tendency to create chloride of lime is of less account in dry light soils than in the colder, heavier soils where the sulphate is undoubtedly preferable, as also for use upon pot plants.

Carbonate of potash is infrequently mentioned among fertilizers, probably on account of what may be commonly considered its excessive cost. There is, however, the counteracting fact that it is a more powerful form of potash, and a small quantity contains a high percentage of plant food. Carbonate of potash, moreover, is caustic, and is capable of destroying much in the way of fungoid and animal life in the soil which may, left unhampered, develop into plant pests. Carbonate of potash is the best of all forms to use in soils which are surcharged with humus and consequent acidity, and nothing better can be used for dusting in potting soil which contains injurious insects or their pupee. It must not be used in large quantities, one pound being ample for three hundredweight of compost. It is well to mix the carbonate of potash with the soil a month before it is used for potting, and when used on seed-beds or special beds where carnations, liliums, or other choice plants which are particularly subject to attacks by ground pests are to be grown a similar period should elapse before sowing or planting. Two ounces per square yard or four pounds per rod is a full dressing for this purpose.

SUPERPHOSPHATES

The extensive use of superphosphate of limer believe the annual output in Britain exceeds one-and-a-half million tons-is sufficient evidence in itself to prove that it has accomplished good service in both agriculture and horticulture, but to merely accept the fact that super phosphate of lime is a good fertilizer is not enough, for quite a large proportion of users are not so well acquainted with the nature of the product, its widely varying quality and consequently widely varying value as it would be well they should make themselves. It is not sufficient that a purchaser of superphosphate of lime should secure two or three quotations and order the cheapest, contenting himself with the thought that he has saved money, for there are low, medium and high grades of the product, and one sample may contain as little as 25 per cent. of soluble phosphates while another contains as much as 80 per cent.

The original method of procuring superphosphate of lime was by dissolving bones with sulphuric acid.

Another method is by converting mineral phosphates into superphosphates, and yet another is by treating both Ehosphates and bone ash with strong Sulphuric acid. So-called ordinary superphosphate of lime is of low grade, containing probably no bone product, and showing a low percentage of soluble phosphates. This should be an inexpensive and useful agricultural fertilizer, but for garden use superior samples which show the highest percentage of soluble phosphates should be used and cannot, despite their higher market price, be considered extravagant, because one buys and pays carriage on the smallest proportion of waste matter.

An erroneous idea exists in some quarters that where lime or chalk abounds superphosphate of lime is not a good fertilizer to use; as a matter of fact it is frequently the particular need of such soil for there may be a serious shortage of phosphates in chalky or limy soils, and it is that shortage alone which superphosphate of lime will rectify. Orchards and soft fruit plantations, especially on light dry soils and on chalk, will benefit greatly by a dressing of about six pounds of medium quality or three to four pounds of high grade per square rod. Superphosphate of lime may also form an ingredient in a mixed fertilizer for flowering plants and general purposes, the other ingredients being potash and something of a nitrogenous character, such as sulphate of ammonia, but not nitrate of soda.

Never use superphosphate on soil which is deficient in lime.

BASIC SUPERPHOSPHATE is the name given to a special brand which is more suitable than the ordinary type for use on clay lands, or on soils which either from lack of lime or an excess of humus are inclined to be sour. Basic superphosphate works at a slower pace than the other forms, and may be used in slightly larger quantities either in autumn or spring. It is a comparatively cheap fertilizer, but the percentage of soluble plant food is smaller, and it should be considered mainly on the ground that it is particularly suited to soils which are more or less sour and therefore not in a condition to benefit by the concentrated acid superphosphates.

LIME

Few substances have been the subject of more pen work and talk than lime, but sometimes one is disposed to believe that lime is really less understood than most things with which horticulturists have to deal.

Lime is not itself a fertilizer, but it is frequently accredited with powers of enriching the soil. Lime of any kind is often declared to be beneficial to any soil, but that is not entirely true. Many limestone rocks contain so large a percentage of magnesia that they are unsuitable for horticultural purposes, for whilst a small proportion of magnesia may be required by certain kinds of plants, an -excessive quantity is actually poisonous to vegetation, and could better be used as a weed-killer than as a fertilizer. Many soils contain a sufficiency of lime, and to add to the quantity in such cases is unnecessary waste and may even be injurious.

Liming a hungry soil will tend to increase its hunger or poverty, and to apply caustic lime to a soil which is deficient in humus may, and probably will, destroy what little humus it contains, greatly to the detriment of the soil.

The kind or form of lime which is best suited for one kind of soil is in all probability unsuitable for soils of widely different character, but often these facts are overlooked or unrealised by those who persistently advocate the use of lime.

The power of lime is great, and its work is many sided. Naturally, if used when its power and activity is not needed, it becomes injurious, for it is very difficult for lime to remain inactive when buried in moist earth. When it is required few things are more beneficial to soil than lime. The one point I wish to emphasize is that it is highly essential that it shall only be used with intelligent care, and should in fact be made the subject of closer study and less glib talk.

Perhaps the greatest real service which lime renders to plant life is the liberation of potash which lies inactive in soils, especially in those of a stiff clayey nature.

The mechanical action of lime in disintegrating the closely compressed particles of clay soils is of great importance, and that part of its work is most thoroughly performed when lime of a caustic nature is used. Caustic or quicklime is made by burning chalk or limestone.

This should be used on vacant land, and should be buried below the surface, but not deeply, because its tendency is to sink or to be washed down by rain. If caustic lime lies upon the surface of the soil it becomes slaked by reason of its power of absorption of moisture from both the soil and the atmosphere. Slaked lime is less powerful than caustic lime. In its ordinary state caustic lime is lumpy and inconvenient for even distribution. That difficulty may be overcome by securing ground lime; but mark the fact that ground limestone is also a commercial article and a totally different thing from ground quicklime.

In this form lime is of greatest service on peat lands, which are naturally almost destitute of lime. It is also the best form for use on old soils which, by long continued cultivation, have become manure sick and sour from excess of humus. Its mission here is to correct acidity, making the soil sweeter and more wholesome.

Slaked lime will perform the same kind of work in the soil as quicklime, but it is milder and works more gently. It is therefore more suitable for soils which are of lighter friable character and not overcharged with humus than for stiff clays and manure sick soils. For very light, hungry soils which are seriously deficient in lime chalk is the best form in which it may be supplied. In the case of chalk it is often beneficial to cover the material over the surface of the soil in autumn and allow the rain to wash it in. There is no danger of reducing the strength of chalk by exposure, because it is not in a caustic state. It is important that lime shall not come in contact with manure or fertilizers while exposed to the atmosphere. Lime immediately attacks the nitrogen in the manure, allowing it to escape into the atmosphere in the form of ammonia. When both lime and manure are to be applied to soil it is wise to dig in the lime and postpone manuring for a month, or as an alternative the manure may be dug in first, and the lime a month or so afterwards.

The important thing is to ascertain whether one's soil requires lime or not. It has been frequently stated that a gardener should never hesitate to use lime, but that is too sweeping an assertion. Testing soil for lime is frequently done by shaking a small sample of soil in a bottle of water, and then adding a small quantity of hydrochloric acid (spirit of salt). If plenty of lime is present effervescence takes place. If lime is present in only moderate quantity effervescence will be weak, and if there is no lime there will be no effervescence. I could never feel quite satisfied that a few ounces or a spadeful of soil put to a test or subjected to analysis may be taken as a reliable means of ascertaining the truth about the whole soil of a garden. A better method of testing soil seems to me to be by opening a series of small holes at various points in the garden and pouring into each a small quantity of the acid. If the soil contains lime there will be no mistake about the effervescence under this method, and it may sometimes be found that one part of the garden has lime and another part little or none.

Ground limestone is suitable material to mix with potting soil, and with the compost for tomatoes, cucumbers and melons. It is also more suitable for both sweet peas and culinary peas, beans and the leguminous tribe of plants generally than is caustic lime. Potatoes also thrive well when ground limestone is used.

The advantage of chalk on hot sandy soils is that it is cooler than other forms of lime and tends to bind the particles together, assisting thereby the retention of moisture.

Gypsum,' or sulphate of lime, is very serviceable as a liberator of potash and also as a fixer of ammonia. For the latter reason gypsum is a good substance to sprinkle over manure heaps when these must necessarily be in the open. It is the only safe form of lime to use for this purpose.

Gas lime is chiefly of service where land is badly infested with insect pests or with rank weeds. It must be on the surface for about six months before being dug in, and cannot be used on land occupied by growing crops. Twenty-eight pounds per square rod should be the maximum amount of gas lime even for the foulest land.

Pounded oyster-shell contains a great deal of lime. It is excellent material for garden soil and also to mix with compost for potting chrysanthemums and other plants.

CHAPTER XV

GREEN MANURING

REFERENCE has already been made in two or three chapters of this book to the value and benefits of what is by common consent termed green manuring, the primary purpose of which is to add humus to soils which are deficient in that essential commodity. Even the simple process of burying waste leaves of vegetables, pea and bean haulm, or weeds is green manuring in so far as it adds to the humus in the soil, but there are certain disadvantages in burying fully grown foliage, stalks and stems. In the first place the tough, fibrous tissues decay extremely slowly when buried to a good depth, whereas they are apt to be a serious source of trouble when but slightly beneath the surface.

Cabbage stumps and hollow stems provide snug retreats for destructive pests, and, further more, there is always a considerable risk that old vegetable refuse may be infested with diseases which will not be destroyed by the simple act of burying them. Woody tissue does not make humus of equal quality to that of tender young foliage, and altogether the advantage is on the side of growing a special crop for the definite purpose of green manuring. The green crop may, under suitable conditions, be a substitute for a dressing of animal manure, but in that case it will be of distinct advantage to help the green crop along with an application of a good, quickly soluble fertilizer.

When a barren sandy soil, deficient in humus, is taken into cultivation it is, as already stated in an earlier chapter, an unprofitable plan to begin by manuring, for as the plant food in the manure is reduced to solubility so it is washed away into the subsoil by the next rain; humus, is required to act as a sponge to hold the dissolved ingredients of the manure. Green manuring will be of immense benefit to such land. Leguminous plants, particularly common lupins, will germinate in very sandy soil, and if assisted with a little soluble fertilizer, the first crop will form sufficient humus to enable a crop of mustard to make considerably more growth. By following this second crop with a dressing of cow or farmyard manure, and immediately sowing tares, the improvement in the condition and texture of the soil will be remarkable, because the tares will take up the plant food as it is rendered available, and having stored its virtue in their leaves and stems, will return it to the soil in an altered but still useful form when the crop is dug or ploughed in.

It is fairly well known that such crops as carrots, parsnips and beetroot, and most bulbous flowers dislike close contact with fresh manure, but still require a well-nourished soil. An excellent method of preparing a plot of poor ground for these subjects is to manure the ground during the summer, sow thickly with rape or mustard, which will feed upon the manure, and when dug in will return a large proportion of that nourishment to the soil in a form more suitable to the aforementioned crops or bulbs. Such method of preparation is commendable to growers of choice daffodils, gladioli, liliums, etc., as well as to cultivators of root vegetables. Various kinds of plants may be used for the purpose of green manuring, but those which make a large quantity of soft herbage in a short time are most serviceable. Lupins are highly esteemed because, in common with other leguminous plants, they encourage the development of nitrogen fixing bacteria or microbes in the soil, and are, moreover, of rapid growth. Tares or vetches are also leguminous, and being plants of a spreading habit which may remain longer on the ground before being turned in, they are of special service on land which is full of noxious weeds. A thickly sown crop of tares, assisted by a dressing of superphosphate and Kainit in equal parts, totalling seven pounds per rod, will do much toward smothering to death many troublesome weeds which are difficult to eradicate by other means.

Mustard is an excellent crop to sow when land is vacated by one crop and will be wanted for another in a month or six weeks time, and this crop has an added advantage in that wireworm and other ground grubs dislike it, and although no claim is made that a green crop of mustard will kill the pests, there is good reason for stating that it will cause them to quit and remain in absence for a lengthy period.

Maize, rape, and even turnip seed may be sown as green crops, but should be turned in before they become coarse and hard. Even some weeds are permissible under certain circumstances; for instance, where a very harsh gravel is too poor for even the above named crops to make reasonable growth camomile has been used with success to pave the way for a crop of tares, the latter smothering any laggard seeds of the weed which may germinate after the main crop has been turned in.

It is scarcely necessary to remark that care must be exercised not to leave the weed crop long enough to allow flowers to open before burying it, lest seed may form and establish an undesirable tenant to the future chagrin of the cultivator. Weeds, as a matter of fact, should only be used in extreme cases when one is compelled to make a garden where no real garden soil exists to form a basis of operations.

These remarks may appropriately lead us on to further consideration of weeds in relation to cultivation of the soil.

The aim of the gardener should always be to secure a weedless garden. Such is seldom if ever obtained, but often failure is due to the methods adopted rather than to lack of industry. The fact is, weeds are not studied sufficiently by the average gardener, either professional or amateur, and by adopting the method of treating all weeds alike much mischief may be done.

Quite a prevalent idea is that the main purpose of a Dutch hoe is to cut up weeds, and it is in consequence considered to be quite unnecessary to hoe a plot or bed until it is covered with weeds. When such hoeing is done in hot, dry weather, the small soft weeds quickly perish, and the plant food they have absorbed from the soil is wasted. Bigger, stronger weeds, however, will frequently retain vitality for some time, and it is not an infrequent occurrence for strong weeds to mature and cast seeds after they have been severed from their roots by the hoe, thus perpetuating the trouble. Some weeds are annuals, and when hoed before they flower or produce seed may be really destroyed, but others are perennial, and in their case merely cutting their tops is tantamount to pruning, for one crown cut will frequently break out and make a much bigger plant with several crowns.

Docks, dandelions, plantains, cow parsley, buttercup and all weeds that make thick fleshy roots of a perennial character require to be eased out of the ground with a fork and removed to the incinerator for burning. It is bad practice to simply throw such weeds on a rubbish heap, for many will grow and produce seeds which may find their way back to the cultivated ground. Obviously it is extravagant to allow such weeds to grow to a large size and then have to remove them, for the nourishment they have taken from the soil is thus wasted. It is likewise folly to bury such weeds, for they are very tenacious of life, and many are quite capable of growing through to the surface even when buried two feet deep.

When ground that has lain derelict for a long time and has become tenanted with rank weeds is taken over for conversion into a garden the first consideration should be to clean it, and no better means has yet been devised than that of forking the soil over, carefully hand-picking every root that can be found. Couch grass and other weeds with stoloniferous roots require much patient labour, because even small broken ends left in the ground are capable of growing into vigorous young plants.

Bellbind and thistles send down their roots to such a depth that they cannot be entirely extricated, and the only hope is that by persistently breaking off young growths before leaves are fully formed the plants will be so weakened that the roots eventually die. No plant can live for a great length of time if it is prevented making foliage. When, therefore, it is impossible through occupation of beds and borders by growing plants to frequently disturb the soil by forking, every endeavour should be made to hoe as soon as young shoots of bell bind or thistles appear.

Should there be at any time thistles which have escaped the hoe and risen to a considerable height, pull them with a steady upward pull, for by this means a good length of root may be drawn, thus prolonging the interval before their reappearance above ground.

In the month of June it is well to leave such thistles as may be above the surface, for if pulled or hoed they will grow away again rapidly. In July, however, just as they are about to open their first flowers their main stems become hollow, and if cut off an inch or two above the ground the plant will grow no more that season. It is imperative, of course, that the cutting shall take place before there is any chance of seed forming. It is maintained by some cultivators that if thistles are cut in July three years in succession the most vigorous crop will die out entirely, but that theory would not encourage me to leave thistles growing until July where opportunity occurs to cut them up with a hoe when an inch above ground.

Where an unplanted plot of ground is covered with nothing but young plants of annual weeds, such as chickweed, groundsel and seedling grass, it is worth while to dig the ground, burying the weeds well under the surface. Such herbage will make green manure, which will compensate for the nourishment the weeds have taken from the soil. No weeds in flower or seed should be buried, nor any perennial weeds. Repetition may be bad form, but the frequency with which this advice is ignored justifies constant reiteration.

By using the hoe frequently before the surface is covered with weeds myriads of germinating seedlings may be killed at birth.

It is highly important that when a site is prepared for sowing lawn grass seeds the levelling and raking of the surface shall be done several weeks before sowing the seeds. This gives an opportunity for weed seeds which are bound to be present in the soil to germinate. They may then be hoed and raked off, leaving a clean bed for the grass seeds.

Every time a plot of ground is dug weed seeds will be brought to the surface, and although they may have lain dormant for many years while deeply buried, they will quickly germinate when brought within influence of air and light. That is one reason why it is an advantage to dig a few weeks in advance of planting or sowing, for it is always easier to hoe unoccupied land than to deal with weeds just as tender young grass or seedling plants are appearing through the soil.

CHAPTER XVI

SOILS FOR POT PLANTS

SUCCESSFUL cultivation of plants in pots depends to a large degree upon the character of the soils used and the method of treatment meted out to them.

Much of the care which the old florists exercised in their selection and preparation of potting soils has been termed "whims and fads" by growers of modern ideas. Some justification there may be for such slighting remark, but there is little doubt it would be safer and better for much plant culture of to-day if plants men should err on the side of excessive caution rather than on the opposite side of carelessness.

Familiarity with the statement that loam, leaf mould, and sand should form the staple ingredients of general potting compost must not misguide the aspiring plants man to the assumption that this constitutes the sum total of essential knowledge of potting composts. One needs in the first place to acquire ability to judge quality and condition of soils, and must learn much regarding their handling and preparation.

LOAM

Many queer soils pass under the name of loam, but the so-called loamy soil of a garden is seldom suitable for potting. It is highly essential that potting loam shall be well supplied with the fibrous roots of grass, and should, in fact, be composed of the top spit of pasture land. The thickness to which it should be cut depends upon the character of the soil in which the grass grows. If this be a fairly dark brown soil, inclining toward clay, but not stiff, stubborn clay of the consistency of brick earth, the thickness of the clods may be as much as six inches. If the soil is harsh, stony gravel, two inches will probably be as thick as will be of any service. Some pastures which have been well kept and well grazed by sheep, cattle or horses will be of good "heart," meaning that the soil is rich and well nourished, and such will make valuable loam.

Some grass lands are extremely poor, and turf from a starved soil is of very little value, the only hope being that it contains sufficient root fibre to make it worth while improving the quality of the loam.

Where most of the grass is dead, and the soil contains very little living root fibre, it is valueless and should be discarded.nor should very young grass which has not formed abundance of root fibre be taken for the formation of potting loam. The tender young roots will quickly rot completely away, leaving only inert close-binding soil.

Turf must not be used for potting when freshly cut; it must be stacked long enough for the herbage to decay and for the root fibre to partially but not completely decompose.

Building a turf stack may be called a simple task, but, like many simple tasks, it requires proper execution and the exercise of a good deal of intelligence.

If the turf is in a comparatively dry state when carted, a deal of the earth will fall away from the roots. A common custom is to build the turves in the form of walls and then throw the loose soil in the middle of the stack. That is an error, for it means that the middle of the heap, where decomposition will proceed at a quicker pace than near the outside, will be occupied by the loose soil, which has least need of decomposition. It is essential, of course, that the walls shall be composed of turf that hangs well together, but because it will remain nearest the air this should be turf which contains most earthy matter. The roughest, coarsest grass and most fibrous of the turf should be placed in the middle of the stack, and as layer by layer is built up the loose soil should be evenly spread over the turves. By this means uniformity of soil and fibre will be maintained throughout the heap.

In building up the walls see that they lean slightly inward as they rise, and at frequent intervals "tie" the turves by placing one endways into the heap. This will prevent slipping when in consequence of decomposition the heap begins to sink.

Turf which is very dry when being stacked should be watered as building proceeds.

Many cultural guides advise the mixing of manure with potting soil at the time of preparation for potting. That, in my opinion, is rarely sound advice. The far better plan is to sandwich between the layers of a loam stack thin layers of either stable, cow or sheep manure, or to sprinkle every layer of turf with poultry or pigeon manure, Such a stack should stand at least a year before use. It may stand two, or even three, years, but at the end of the first year a good waterproof thatch should be placed over the stack to keep out excessive rains. It may be necessary to lime a loam stack. This point may be ascertained by applying the spirit-of-salts test to a few slabs of the turf (on the underside) at the time of building, and in the event of lime deficiency being detected the best plan is to cover alternate layers with lime instead of manure; or gypsum instead of ground lime may be used with manure over every layer, a mere sprinkling being used each time.

When the loam has sufficiently decomposed: the stack should be opened at the side, and a quantity of the loam removed from the centre of the stack. Chop this down with the spade and spread for awhile on a hard floor in the open, where it may have the benefit of the sweetening influence of the air and, if possible, of frost.

To this point I attach great importance, yet it is commonly neglected, and is rarely insisted upon by writers. A moment's thought should suffice to show the necessity for this exposure of loam from stacks. It is common knowledge that soil which is buried beyond reach of air quickly becomes sour. Add to that fact the further knowledge that decomposition of the blades and roots of the grass will cause a certain amount of acidity, and then ask the question why gardeners, who will lay great stress upon the importance of autumn digging and of aeration by surface hoeing in summer, will carve their way into the centre of a big turf heap and take loam which has lain a year or two tightly locked in several feet from the surface or sides of the stack, to use it immediately for potting young plants. Does that not appear to be a strange folly?

The ideal arrangement will be to transfer the loam, after it has been frozen through or soaked by one good downpour, to an open fronted shed where it may lie loose and in shallow heaps. From this store potting composts can be made as required, and from time to time the diminishing quantity will be replenished from the stack. Loam thus treated, enriched by the addition of manure when stacked, assisted when necessary by a judicious sprinkling of lime, and allowed the benefits of aeration before use will be capable of growing anything, and the enhanced value has been secured at the expense of just a little additional care.

Such loam requires only to be chopped down with a spade, except when to be used for quite small pots or for seed sowing, when, of course, sifting is necessary. It is, however, a frequent occurrence for the sieve to be used and the fibrous substance which should be left in the compost to be extracted and thrown away. That practice very probably will be the first step toward creating trouble with watering, for the fibre is essential to maintenance of a proper degree of porosity.

LEAF MOULD

Whilst readily acknowledging the utility of good leaf mould, and admitting that its use as an ingredient for potting soils is often a real necessity, I am of the opinion that it is frequently overrated, and that at any rate it is a mistaken notion to suppose that any sort of leaf mould in any condition is capable of improving the quality of potting soil. There are vast differences in leaf moulds, which used to be taken into account. Some are composed of leaves which contain far too much woody tissue, and many samples of leaf mould are injurious by reason of poisonous acids they contain or of fungoid diseases which flourish in them. Humus is the chief element of value in leaf mould. Humus should form about

10 per cent. of a well-balanced soil. Some good samples of fibrous turf loam with which a reasonable quantity of manure has been incorporated will in themselves contain 10 per cent. or thereabouts of humus, but one may see prescriptions for making up potting composts which stipulate for equal parts loam and leaf mould with an addition of sand. Such a mixture may easily contain forty to fifty per cent. of humus, a quantity excessive and for some plants positively injurious, even allowing for the fact that in the confined area of a plant pot a much larger proportion of humus is beneficial and essential than for similar plants in the open ground. The best leaf mould is that of oak or beech leaves which have fallen from the trees in a wood or plantation and have gradually decayed in a perfectly natural manner. Such leaf mould may sometimes be raked or scraped off the ground under the trees where it lies in a loose soft mulch to a depth of several inches. It is sweet, wholesome, and precisely what is required.

The worst kinds of leaf mould are made by dumping all manner of leaves, hedge clippings and kitchen garden garbage into a hole where they become saturated, fermented, and converted into a putrid, mucky condition.

When leaf mould is to be "made," the first thing should be to prepare a hard, flat base, where surplus water will quickly drain away. The leaves should be of the soft kinds of tree leaves such as English oak or beech, and should never contain those of laurel, holly, evergreen oak, or other thick, hard and leathery kinds.

When a load of leaves is dumped it should be spread evenly, saturated with water, trampled firmly, and if possible covered with a thin layer of fresh stable manure. When a good-sized stack has been built up in this manner it requires at least a year to decompose; and when at length it is to be used access should be gained to the middle of the stack, a one-inch-mesh sieve should be used, and any rough undecayed leaves which do not pass through the sieve should be returned to the bottom of the stack and covered.

The leaf mould will also require exposure to air, for it is in all probability too sour for immediate use.

The proportion of leaf mould to be mixed with loam for potting depends to some extent upon the kind of plants to be potted, but it depends even more upon the character of the loam with which it is to be mixed. If the latter contains abundant fibre and some decomposed manure, a small addition of leaf mould will suffice. If the loam is of a heavy character, and comparatively short of fibre, a large proportion of leaf mould may with advantage be added; but a common failing is to use more leaf mould than most plants require. For carnations none is needed, so long as the loam is friable and fibrous, but for young seedlings of quick, soft growth leaf mould in liberal proportion is useful.

OTHER INGREDIENTS OF POTTING SOIL

Sand is, with very few exceptions, a necessity in potting composts. Some loams, of course, are of a gritty or sandy nature, and the addition of further sand will be moderated accordingly. Where, however, the loam is inclined to be heavy and the leaf mould soft, add sand with a liberal hand, but it must be clean, sharp, and coarse grade silver sand. The soft sands which builders use are useless for potting soil. The builder requires a sand that binds; the purpose of sand in potting composts is to maintain porosity, and for this purpose there is nothing to compare with the Bedfordshire silver sands.

To prescribe certain proportions of the various ingredients of a potting compost without knowing their respective characters is not calculated to ensure satisfactory results; the addition of sand, etc., must be dependent upon the character and condition of the other ingredients.

Charcoal is capable of keeping soil in a sweet, wholesome condition, and is especially valuable where plants require an abundance of water. It may be used in the form of fairly large pieces, in which case it should be placed separately in the pots upon the drainage crocks, or it may be used in the form of finely broken pieces about the size of wheat grains or even as large as peas. In this form it may be mixed with the compost, a five-inch potful being sufficient for a bushel of soil.

For seed pans powdered charcoal may be lightly dusted over the surface after sowing the seeds. It will thus tend to check the development of moss and liver-wort.

Burnt earth, old pounded lime rubble, pulverized oyster-shell, all serve a useful purpose when composts are mixed for potting rock plants or for sowing their seeds. Form 5 to 10 per cent. of the whole bulk of compost may be composed of these ingredients separately or in mixture.

Sphagnum moss cut up into small pieces is used with advantage when seeds or seedlings of bog plants, aquatics, terrestrial orchids and some few of the moisture loving primulas, etc., are to be potted, and these, as well as ferns, azaleas, and many warm greenhouse plants, require peat instead of leaf mould.

For most ordinary purposes the sandy peats of the moorland are most serviceable, but orchid growers use a loose, coarse, fibrous peat that is free from sand and earthy matter.

Peat when required for potting needs no other treatment than to be pulled to pieces of convenient size, according to the sizes of the pots to be used. It should not be mixed with manures of any kind; sometimes it is used in conjunction with very fibrous loam, but it is seldom advantageous to mix either leaf mould or sand with a good sample of peat that is to be used for potting real peat loving subjects.

Bone meal, guano, and other fertilizers are

sometimes mixed with potting soils. It is quite a legitimate practice where strong growing plants such as chrysanthemums, hydrangeas, pelargoniums, etc., are to be transferred to their flowering pots, but for younger plants it is imprudent to use strong stimulants.

So far as animal manures are concerned, it has already been recommended that these may be sandwiched between the layers of a loam stack to decompose simultaneously with the vegetation of the turf. That is an infinitely better method than adding manure with the other ingredients during the mixing of composts for immediate use.

OLD OR SECOND-HAND POTTING SOIL

Much potting compost is used for growth of plants which occupy the soil for one brief season, and the question sometimes arises what should be the fate of this second-hand potting soil. It is obviously too good to waste, but at the same time it is very often in an unfit condition to be immediately re-used, and it is seldom wise to use it a second time for precisely the same purpose; During a whole season's growth a large amount of water has been poured into and passed through the soil, and with it nourishment has disappeared, that wastage being additional to the absorption of nourishment by the plants. The fibre and humus in the compost has deteriorated and the soil will have become hardened and solidified, although probably this is counteracted to some extent by the amount of root the plant has left in the soil. It is sometimes considered ample for all requirements if the old soil is mixed with some fresh loam and leaf mould, but the net result of that course may be to bring the new soil into the same unsatisfactory condition as the old, for sourness and acidity is more capable of spreading than of being beneficially influenced by mere contact with good soil. Fresh air, acting upon old soil spread loosely in the open, will do much toward sweetening it, and a dressing of powdered. lime (freshly slaked) will further correct acidity. After a period of such recuperative treatment it remains only to see that any necessary replenishment of fibrous material is attended to and the soil will be quite fit for re-potting. There will be an advantage in the case of a quantity of compost that previously grew a particular kind of plant in using it for some entirely different class, but where the previous occupants were of a mixed, miscellaneous character, such a point of consideration will not arise.

A further matter of considerable importance concerning the handling of potting soils remains to be dealt with, i.e., the degree of moisture it contains at the time of handling. It is a well established fact that soil should not be dug or otherwise disturbed when in a saturated condition, but it is not always realized that potting soil should not be handled when excessively wet. Where the advice is followed regarding a roofed shelter for soil which is to be used for potting there will seldom be trouble from too much moisture, but when the only storage available is in the open air it will very often be necessary to bring in the supply to the potting shed and leave it in a loose heap for a sufficient length of time to bring it to a workable condition.

In the middle of summer one may frequently find both loam and leaf mould extremely dry; a state in which it is quite impossible to pot properly.

There is a right and wrong way of moistening such compost, the wrong way being to pour into the heap a large volume of water that will reduce the soil to a muddy condition. The proper method is to place the heap on a clean, hard floor, and starting at one end proceed to turn over the compost with a shovel, continuously sprinkling with water through a fine-rosed can as the turning proceeds. If the water for this purpose can be boiling so much the better; the scalding will destroy disease germs or insect pests that may be present in the loam or leaf mould.

After sprinkling in this manner leave the heap for a night and turn again next day.

The correct condition for potting can best be ascertained by squeezing a handful of the compost in the hand; if it binds, and is with difficulty made to crumble again, it is too wet; if it remains quite loose and disintegrated after pressure, it is too dry; but if it yields to pressure and can still be crumbled by a slight tap, it is in excellent workable condition, and rest assured it is practically impossible to pot really well when soil is not in such condition.

Soil sterilization by steam or by roasting is strongly advocated by some growers of choice pot plants as a safeguard against both fungoid diseases and insect pests. Probably it is a wise course to adopt where old soil has to be used for fleshy rooted subjects, and especially for very fine seeds. Steam, however, may do harm as well as good to soil, and roasting is calculated to rob humus of its utility. Generally speaking it should be adequate preparation if pans in which such seeds as begonia and gloxinia are to be sown are scalded with boiling water the day previous to sowing.

CHAPTER XVII

SOIL TREATMENT IN PERMANENTLY PLANTED QUARTERS OF GARDENS

APART from those quarters which are periodically rendered vacant and can consequently be dug, manured, limed and improved by aeration, there are in most gardens certain plots, beds or borders which are permanently planted, and these must not be denied a certain amount of attention.

Many people seem to imagine that a shrubbery can look after itself, and never requires manuring. The rooting area of hedges is frequently neglected, and the thin, bare bases, gaps and weak places in garden hedges are frequently the direct outcome of neglected soil.

Rhubarb beds, raspberry plantations, herb borders, asparagus beds are portions of the kitchen garden that are frequently neglected, and many mistakes are made in the treatment of the soil in which roses, preonies, rhododendrons, lily of the valley and other flowering plants of a more or less permanent character are grown.

In some cases it is possible to loosen the surface soil with a fork without injury to the roots of the occupants, as, for instance, between strong growing shrubs such as lilacs, berberis, ribes and many others. The soil may usually be forked up alongside hedges, affording opportunity for removal of rank weeds, and wherever this is possible it should be done, and occasionally manure of some kind should be worked in during the digging, for even hedge plants and shrubs require nourishment.

It is not, however, a good plan to fork up and disturb the roots of rhododendrons, neither do raspberries benefit by having their roots torn and lacerated. The point is that both require a certain degree of aeration as well as nourishment of the soil, and in the case of raspberries this need is the greater by reason of the amount of treading the soil endures during the gathering of the crop. To minimise this evil it is important to mulch the ground between the rows of raspberries prior to the ripening of the fruit, using material of a loose, springy nature, which will minimise the compression of the soil under the tread of feet. The material may be stable manure with an adequate quantity of long straw, laid on several inches thick. Take the precaution to see that the ground is well soaked before putting on the mulch, for a loose mulch overlying dry soil is apt to soak up the bulk of any rain that falls, leaving the soil beneath still dry; but if the ground be wet, evaporation will keep the mulch tolerably moist, with the result that it will allow more of the rain to pass through to the benefit of the soil and roots. Gorse or heather makes good resistant mulching material for the purpose, and has the great advantage of allowing free access of air to the soil.

After the crop has been gathered and the old canes and superfluous suckers have been removed, the remains of the mulch should be cleared away, and the next thing should be to make a series of, holes between the rows of canes, using some such implement as an ordinary household fire-poker: for the purpose. By thrusting the poker vertically into the soil to a depth of about six inches at intervals of a foot or thereabouts access of air is provided with a minimum of injury to the roots. The holes may be left open until they close involuntarily by the washing of grit into them by rains. The effect of such aeration of the soil is quite remarkable.

Dressings of fertilizers may be occasionally required, but with the manure mulch every season such may be less essential than a dusting of lime over the entire surface, say, every third year. The length of time a raspberry plantation remains prolific and profitable depends upon the culture it receives, but such treatment as is here described is capable of considerably extending that period. The perforation of the soil with a poker may be practised with advantage on rhododendron beds, but where lily bulbs are planted between the shrubs (an excellent situation for many fine liliums) care must necessarily be exercised to avoid piercing the bulbs. Sheep manure, or Wellson's Plant Food is of special benefit to rhododendrons, but it is only necessary to apply manure when growth and flowers reveal signs of starvation.

The mistake is often made by amateurs of piling up manure close around the stems of rose bushes. Such a practice is the reverse of beneficial. A Mulch of short manure spread evenly on the soil in autumn may be of great service, but a good six inches should separate the manure from any stem. In spring the remains of the mulch may be lightly and very carefully forked in, using an old much-worn fork, and taking care not to punish the root. It is not essential to manure rose-beds every year. A slight sprinkling of ground lime the second year and the third year about two ounces of fine bone-meal per square yard will keep an average rose bed going comfortably, but the bone-meal should be forked or hoed in, not left on the surface, for it will not decompose. while left uncovered. Peruvian guano or a proprietary brand of rose fertilizer may with advantage be used as a change. When growing on poor, light soil it may be necessary to apply animal manure annually, but continual surface dressings on beds that cannot be deeply dug, and sweetened by periodical exposure to frost, tend to produce acidity of the soil. Paeonies and other plants which should not be disturbed when once planted should have the surface soil very frequently hoed. The perforation as recommended for raspberries may also be adopted with advantage, and nourishment should be afforded alternatively by mulching, fertilizers, and liquid manure.

Rhubarb plantations may be treated in an exactly corresponding manner.

Asparagus beds should not be dug. The surface may be pricked up very carefully with a fork, and a mulch of short stable manure may be spread immediately cutting has finished. Occasionally seaweed may be used with advantage, but it should be covered with lawn mowings or other light litter, such as spent hops, otherwise it will dry up so rapidly that little benefit will accrue. Salt is everywhere spoken of as a valuable fertilizer for asparagus, and in small doses it will improve the quality of the crop. It is, however, frequently used in larger quantities than it should be; one ounce to the square yard being an ample dressing, and that should not be given annually, but once in two or three years.

The effect of salt upon soil is to increase its capability of holding moisture. Whilst, therefore, it is of benefit to porous sandy soils, it is capable of aggravating the defects of cold retentive clay. The lawn is another feature of a permanent character which must be the subject of watchful care.

Dressings of fertilizers should be given annually, because constant mowing means that nourishment is continually being taken from the soil, none of which returns in the form of decayed foliage, etc. The character of the soil should govern the precise composition of the fertilizer, but good proprietary lawn manures are serviceable for general purposes. Guano is good for lawns, and so also is a mixture of basic slag and Kainit, but basic slag is credited with ability to encourage development of clover, and is disliked by some people on that account.

Whatever is used in the way of fertilizer should be mixed before distribution with at least four times its bulk of a "carrying" material, and advantage of this fact may be taken to improve the texture of imperfect lawn-bases. A stiff, heavy soil, for instance, or a wet spongy soil will steadily improve if the annual dressing of fertilizer is mixed with clean sharp sand. A light, sandy soil which is deficient in humus suggests the use of sifted leaf-mould or spent hops as a "carrier" for fertilizers, and lawns on soil of good quality will be well served by a mixture of sifted soil from cucumber beds or second-hand potting soil and superphosphates or other chemicals.

Powdered charcoal and pulverised burnt earth make a dressing that will check development of moss on a sour soil, and as poverty of soil is equally responsible with defective drainage for mossgrowth on a lawn, it is well to mix with the charcoal and burnt earth something rich in the way of plant food, such, for instance, as a good Peruvian guano.

Perforation of turf is a very common need for restoration of health when the soil of a lawn has become hidebound through excessive rolling of trampling.

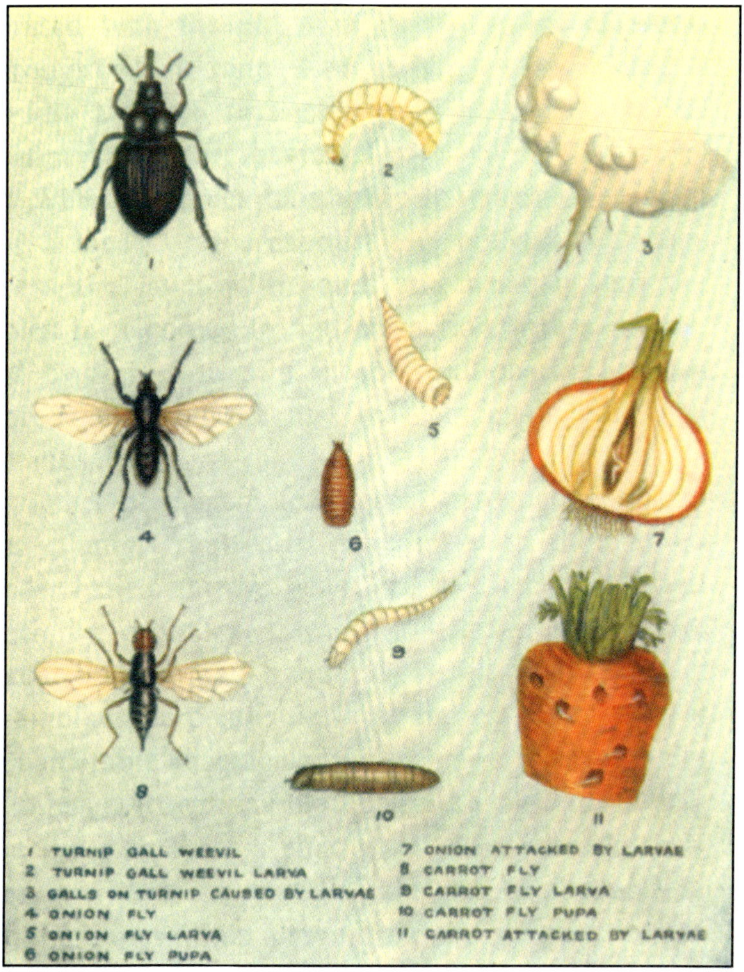

VEGETARIAN INSECTS THAT INFEST GARDEN SOILS

A patent spiked roller is made for the purpose of perforating sports grounds and other large areas of turf, but small grass plots can be well Perforated with the aid of a heavy square of wood through which long wire nails are driven, and a handle attached to enable it to be used after the manner of a turf beater.

When the holes thus made have remained open for a week or thereabout it is advisable to top-dress the lawn with sand, and work it into the holes by vigorous to and fro action with a broom, for by thus filling in the perforations with sharp porous material the turf will not again condense so closely.

Other permanent features of gardens are climbing plants, frequently growing with their roots under hard pathways where the soil cannot be dealt with by ordinary methods of top-dressing, etc. Great benefit may frequently result from sinking drainpipes perpendicularly at intervals, covering a considerable distance from the main stems of the climber. . Rough rubble should be placed beneath the pipes to form a sink away, otherwise the tendency will be to "puddle" the soil at the bottom of the pipes, thus preventing the distribution of moisture among the roots. Liquid manure may be poured into these pipes to soak through to the root run of the plants, thus keeping them well supplied with nourishment.

CHAPTER XVIII

INSECT AND OTHER PESTS IN PLANTED SOILS

ANOTHER matter that concerns planted areas of various kinds is the eradication of soil pests.

Drastic measures may be taken and are justifiable when dealing with vacant land that will not be planted for several months, but it is a different matter to deal with slugs, wireworm, woodlice and millipedes in a bed of choice pansies, carnations, tender young seedlings, etc. Here some caution is necessary, for some of the most powerful substances that may be used on bare ground will do harm to tender young plants. It is unnecessary to repeat what has already been written regarding gas lime and other articles of use, but it may be remarked that for beds and borders of strong, well-established plants the majority of advertised proprietary soil fumigants are safe if the method of application adopted is that of boring holes with a dibber between rows or clumps of plants, putting a spoonful of the chemical into each hole and immediately closing in the soil.

Old weathered soot is a good protector for

seedling plants of most kinds, and Limbux, which is an exceedingly fine grade of powdered lime freed from impurities, is also a deterrent to slugs. One should endeavour to get the surface of the soil as completely covered as possible with as little of the powder as possible on the leaves and stems of the plants. This latter remark is made despite the frequency with which one may read that seedling plants should be dusted with either soot or lime. It is never good to choke the respiratory pores of plant leaves, and that method of combating slugs, etc., should be only a lastesort when utter destruction is otherwise threatened by a veritable plague of the pests. Slugs and woodlice, leather-jacket grubs and millipedes are night feeders, and may be discovered by the aid of a light after dusk. There is still a good deal to be said in favour of making raids upon the pests, picking off and destroying those that are actually on the plants. Those seen crawling on the ground can be killed by dusting their bodies with spent calcium carbide, with powdered lime, or with alum in powder form or dissolved in water. Frequent use of the Dutch hoe does its share toward hustling ground pests from their retreats, exposing the larvae and pupae to insectivorous birds which will atone for taking toll of ripe fruit by ridding a garden of a vast number of foes.

When vegetable quarters are infested with some pest which is particularly fond of one certain kind of crop it will be prudent not only to dress the ground with the best available remedy, but also to give the land a rest from that particular crop and anything of a kindred nature. Thus, if onion-fly attacks a crop the application of favourite remedies such as soot, salt, sulphur, or a proprietary soil fumigant should not be considered ample security for next season's crop, and not only should a fresh plot of ground be selected for onions, but neither seedling leeks nor seedling shallots should be grown on that bed for a year or two.

It must be borne in mind also that pests which attack cabbages, broccoli, etc., may also attack turnips, the gall weevil *(Centorhynchus sulcicollis)* being an example; therefore, after fumigating the soil let some totally different crop occupy it for a year or two before again trying either green crops or turnips.

In the flower garden diseased foliage and stems should be cut away and burned, not left to lie on the ground and infect the soil. Surface hoeing should be looked upon as a means of dislodging insects and their eggs, for plant pests that seek cover under the surface soil will not long suffer the discomfort of continual or frequent disturbance.

ANTS IN THE GARDEN

Ants are frequently a nuisance in flower-beds, seedling beds or in lawns, and by their constant disturbance of the soil in forming ant-hills they dislodge roots and cause plants much distress. Furthermore, ants encourage and protect aphides, and are consequently undesirable tenants of rose beds, etc.

The old-fashioned remedies of paraffin oil, boiling water, etc., are unsuitable weapons to us among growing plants, but in the majority of cases, except in beds where very young seedlings are growing, carbon bi-sulphide may generally be used with success and without damage to plants.

Carbon bi-sulphide is an evil smelling liquid which evaporates in the form of a highly inflammable gas, and it is important that no lighted pipe, cigarette or fire of any kind shall be brought near: while the chemical is being applied. The method of application is to pierce a hole three or four inches deep with a dibber, as close as possible to the ants nest; pour in a spoonful of carbon bisulphide and immediately cover in the hole with soil. The fumes will permeate the soil around and asphyxiate the ants.

Where ants infest seed beds, frames, or green houses, a good plan is to dust castor sugar into an old sponge, and place it near the haunts of the ants. It will soon be swarming with the little creatures, feeding on the sugar. Drop the sponge into a pail of boiling water; rinse out the carcases, and after drying the sponge repeat the process.

MOLES IN CULTIVATED SOILS

Much trouble is caused by moles in gardens, and that the trouble is of far-reaching character may be gathered from a glance at the illustration, which shows very clearly, after patient and painstaking investigation, the actual character of the tunnels made beneath the surface of the soil, and at a

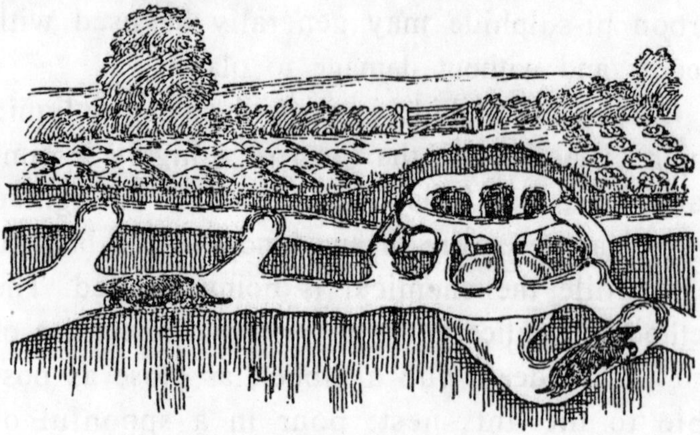

THE HOME AND UNDERGROUND WORKINGS OF MOLES

depth which interferes seriously with a very large proportion of the plant roots in an average garden. It is true moles create these tunnels in order to pursue the ground insects such as grubs and slugs, upon which they feed and which are themselves enemies of the garden, but in this case the enemy of our enemy is a greater menace than the objects of their pursuit, and it may be added that it will usually be found easier as well as more profitable to use means of destroying or of driving away the insect pests than to combat moles when once they discover in a garden a happy hunting ground. Moreover, moles devour large quantities of common earthworms, and although owners of lawns are disposed to look upon worms as a nuisance, there is, as is shown elsewhere, a good deal to be said on behalf of the earth-worm. Apart from the grievous damage moles may do to the roots of growing plants, the tunnels, as our drawing shows, are of a character and dimensions that will, in a dry season, rob the surrounding soil of a great deal of moisture by allowing it to collect in the tunnels and get clear away from the upper soil here it is required. Strange it seems that no ready and reliable means of destroying moles has been invented other than the irksome and somewhat difficult method of trapping. In an old volume, entitled *Systema Agriculturae,* dated 1681, I find the construction and setting of a mole trap explained, which is just as workmanlike and efficient as the best thing available to-day; and the further information is given that freshly burnt garlic dropped into a mole run and covered over will drive moles from their run, and that if the carcase of one dead mole be placed in a run, the living moles will forsake the place for ever.

These hints may prove serviceable where a run happens to be situated close to some special plants, but the surest method of ridding a whole garden of the pests is still apparently that of setting traps in the main runs, and the method of setting is more easily learned from demonstration by a countryman of experience than by written instructions. The chemists who have provided us with so many insecticides and fungicides might well devote some effort to the discovery of some simple and sure means of painless destruction of moles, or at any rate something that will prevent their working in soils where particular crops are planted. I have no original method of extermination to offer, but have one word of warning to utter. Whenever a mole-course is discovered close beside growing plants the temptation is to tread the soil back to its normal condition. This does more harm than good, because the mole will simply re-open its run and repeated disturbance of the soil about plant roots is worse even than a permanent cavity. It is only when the mole has been destroyed that the soil may with safety be trodden or made firm.

CHAPTER XIX

SOME ODDMENTS OF INFORMATION

THERE are many small points the beginner in horticulture will need to learn about soils and fertilizers, and these few notes, if studied and committed to mind, may be the means of avoiding errors and will make the pursuit of gardening more fascinating.

The initial steps to be taken in soil improvement should be the same, whatever the plants or crops to be eventually grown may be; stubborn clay must be broken down, lightened, provided with humus and lime, and in extremely bad cases opened up by incorporation of shingle, ashes or burnt earth.

Dry sandy soils should be improved by incorporation of clay as well as humus, and both will benefit by applications of manure in accordance with the instructions given in previous chapters.

When, however, soil is in fairly good condition, it is by no means necessary or wise to add manure for every crop.

Beds on which seeds are to be sown (for transplanting) should not be freshly manured; a stiff, sturdy growth of seedlings is better than soft, sappy growth, such as will be induced by fresh manure.

Vegetable root crops should not be grown on freshly manured soil, as the result will be coarse, rough roots, forked and misshapen. Until the subsoil has been brought into good condition globe beet, short-horn carrots and stump-rooted parsnips will be more successful than the long, tap-rooted varieties.

The best method of arranging situations for root crops is to grow on quarters occupied the previous season by such crops as celery, onions, spinach, or greencrops, for which the soil was liberally manured.

Although celery is ravenous for nourishment and must be grown in very rich soil, it is not a good practice to place a solid layer of manure several inches thick in the bottom of the trench. No roots of plants can make normal progress when embedded in solid masses of manure. Mix soil and manure as thoroughly as possible and cover with a sufficiency of light soil to give the roots a start before they come in contact with the richer compost.

When planting potatoes do not place the sets immediately upon manure. The old idea that by concentrating manure in trenches immediately under the tubers the plants will reap the maximum of benefit is erroneous. When the manure is spread evenly through the whole area of soil the roots are encouraged to spread themselves in

A Cotton Plant.(210a)

The same Cotton Plant.(210b)

A Catch Crop of Tomatoes under Grapes.(211)

search of nourishment, and by thus providing the young tubers with plenty of room greater development is secured.

Bulbous plants do not like fresh manure. The method which will ensure best results is to manure in spring; grow a green-manure crop which may be dug in during summer, and plant the bulbs in early autumn.

The general rule with strong, quick-acting fertilizers and also with liquid manure should be to commence feeding only when plants show flower spikes, and with such vegetables as peas, beans, cucumbers and vegetable marrows the time to start feeding is when the first pods or fruits have set. Tomatoes should never be fed before they have set fruit, neither should they be grown in very rich soil, or the result will be thick, sappy stems, big, coarse leaves, and very little fruit.

When a well balanced manure or fertilizer is used which creates all necessary forms of plant food, any plants or crops will be well nourished, but when a chemical manure is used which provides only nitrogen, or potash, or phosphates, plants which do not require that particular food in large quantities will fall sick. It should be remembered that plants take up their food in liquid form; they do not absorb solid matter. They must, however, absorb moisture, and will take up whatever solutions may be in the moisture at hand, for plants have not discretionary or selective power which will enable them to take up what is beneficial and reject what is Injurious.

As a general principle, therefore, it is prudent to use a variety of manurial substances which are calculated to maintain a fairly even balance of the essential plant foods. At the same time there are some kinds of plants which require a greater proportion of some one kind of plant food, such as potash or nitrogen, and which are therefore to be benefited by fertilizers which have a high percentage of their particular requirement.

There is justification in this fact for the practice of some manufacturers of offering special brands of fertilizers under such names as rose manure, sweet pea manure, lawn manure, tomato or potato manure; and while such specially prepared fertilizers are available it is wiser for the average amateur to take advantage of them than to endeavour to provide his plants with their particular needs by using crude chemicals.

Soot is by many people credited with far more nutrient value than it really possesses. The only plant food soot contains is a small and variable amount of sulphate of ammonia-never so much as a pound in a peck of soot. The frequent assertion that soot-water is a valuable liquid manure is therefore based on imagination rather than fact. It simply serves the purpose of a weak solution of sulphate of ammonia.

Soot in a dry state, however, is useful as a top-dressing to beds of plants with young, succulent foliage, for it will not only supply an appreciable amount of nitrogen (in the form of ammonia) which will promote healthy growth in young plants, but it tends to check the advances of undesirable insect pests. When freshly made the carbon in soot is capable of injuring young foliage, and on that account old soot is safer to use. It should, however, be stored under cover, for if exposed to much rain the ammonia will be washed out and its food value nullified.

The ignorance of chemistry on the part of many would-be teachers of horticulture is evidenced by the frequency with which the recommendation has been made that a mixture of soot and lime should be sprinkled around plants.

Immediately lime touches soot it attacks and dissipates the ammonia, which escapes into the atmosphere in the form of a strong smelling gas, leaving only a black, valueless substance on the ground.

Soot may, nevertheless, be used with advantage on chalky soil if it is well hoed in directly it has been spread.

BACTERIA AND NITRIFICATION

There is small doubt in my mind that if this book were allowed to go to press without some reference to nitrifying bacteria, or, as some have it, "the manufacture of nitric acid by microbes in the soil," some critic would pounce upon the omission as being a grievous shortcoming.

In the early years of the present century a new doctrine was preached with much vigour, and was hailed as a startling discovery that would revolutionise our ideas regarding soil improvement and alter all our methods of providing plants and crops with essential nourishment. Great emphasis was laid upon the fact that when vegetable matter is buried in the soil it remains a useless encumbrance until it is attacked by millions of bacteria which break down the tough tissues of the leaves and stems and convert the organic nitrogen they contain into available plant food.

That in itself was quite an interesting item of scientific information, but it was eagerly caught up by various theorists and dabblers in profundities, who between them made strange tales.

By some it was contended that the breeding of this mysterious manufacturer of nitrogen in the soil was of greater importance and benefit than the old-fashioned, out of date methods of digging manure into the earth. Pamphlets were published in which it was declared that manure is both unnecessary and useless; articles were written advbcating the idea of breeding and carefully nursing the nitrifying bacteria, and it was even hinted that careless, unscientific methods of soil cultivation might murder the beneficient microbes with dire results. Bacteria cultures on gelatine in glass tubes were offered as the great means of securing prodigious fertility in hitherto barren soils, and bacterised peat dust was brought forward as the last word in soil fertilizers. It was ere long announced that not one but many kinds of bacteria exist in the soil, each playing a different part in the task of creating plant food out of decomposing vegetable matter, and that others are capable of performing similar miracles by attacking carcases of buried animals, bone, skin, fat and feathers. One set of microbes are said to change organic nitrogen into ammonia; a second group changes ammonia into nitrous acid; and a third turns nitrous acid into nitric acid. Nitric acid uniting with lime becomes nitrate of lime, whilst that which combines with potash becomes nitrate of potash, both of which can be taken up by growing plants.

Another idea was brought forward by an enterprising firm who declared that all these essential bacteria exist in plenty in the soil and that the one necessity is to provide them with palatable food, the particular form offered being peat dust. That in plain words brings us back to the old truth that what poor soils require is humus forming material which when bured decomposes as a result of the action of soil bacteria. The additional fact of importance is that humus alone is insufficient, but that the soil requires lime, potash, etc., to complete the manufacture of available plant food.

The story of the work of nitrifying bacteria is interesting, but the knowledge of that work has not brought us to the stage when we may profitably abandon the practice of manuring land, nor has it been proved necessary that the horticulturist must set up a breeding establishment for providing his plants with flocks and herds of invisible microbes or bacteria that shall keep their larder stocked with staple foods and delicacies.

The day may come when a farmer or gardener may carry in his waistcoat pocket a phial or tabloid which will be of equivalent value to his whole stock of natural manures, but that day is not yet; when it arrives this book may become an obsolete but interesting relic of an old-fashioned age.

EARTH WORMS

Common worms have a great influence upon soils, and it is to be feared that much misconception prevails in regard to their real purpose and utility, for it is not unusual to hear earth worms spoken of as pests or to be asked what can be done to destroy or drive away worms. Owners of lawns and players of tennis, golf, and bowls are particularly antagonistic to earth worms, but whilst it is readily understood that the presence of innumerable earth-casts on a bowling-green, putting green, or tennis-court is a nuisance to players, there is considerable danger in adopting measures of wholesale destruction of worms, and many lawns which have been treated with "worm killers" have subsequently caused much anxiety through their sickly and unhealthy appearance.

The notion that worms are detrimental has doubtless been encouraged by the persistency with which growers of pot-plants urge that worms in the pot do harm, but there is a vast difference between the restricted root-run of a plant in a pot and extensive area of soil in a garden, and it is necessary that this fact shall be kept in mind

WORM TUNNELS IN SOIL

A. Wormcast note leaf at B drawn underground by worm..

when tempted to argue that if one worm in a pot is capable of working mischief a large number of worms in a plot of land will do a proportionate amount of harm.

Worms create narrow tubular tunnels in the earth, which descend from the surface to varying depth. It is partly by consumption of the soil itself that the worm contrives to make the tunnels through which it may travel, but only a minute proportion, if any, of the actual soil is digested, the greater proportion being regurgitated around the mouth or entrance of the tunnel in what are commonly called, worm-casts. The real nourishment of the worm is found in rotting leaves and vegetable matter and in the carcases of other worms, for they are cannibals in so far as they will feed on the mutilated carcases of their species. Worms will draw into their holes bits of leaves (see illustration) and will devour a considerable proportion of such leaves, and analysis has proved that worm-casts contain a higher percentage of plant-food in a readily soluble form than the soil upon which they lie.

The worm casts are expanded by a kind of fermentation, and when dried on the surface, but not baked by hot sun, they will pulverise more readily than the normal soil. Therefore by sweeping a lawn vigorously after dew has dispersed, the worm casts may be evenly distributed with considerable benefit to the surface roots of fine grasses.

The tunnels themselves are of much service, for they give access of air into the soil to a greater depth than would otherwise be available, and as so frequently reiterated throughout these pages, without air there can be no fertility in the soil; no healthy root activity.

During heavy rains the worm tunnels serve as channels for the rapid passage of water to the subsoil, thereby improving drainage and preventing the top soil becoming chilled with stagnant water. With these benefits in view it must be apparent that it is disadvantageous to rid a plot of ground of worms, and the more so in the case of lawns and greens, which cannot be periodically aerated by digging, and where frequent rolling and treading tends to consolidate the soil. Imperfect drainage is also highly detrimental to the health of grass, and altogether it is a grievous misfortune to a lawn to lose its worms. Far better would it be to spend more time in sweeping the grass; this operation in itself is more beneficial than continuous rolling, and it would remove the sale cause of objection to the worms whilst retaining the benefits of their untiring activity.

The differences between the worm in the open ground and the worm in a plant pot are that in the confinement of a pot a worm will create so many tunnels that water escapes through them without soaking through the rest of the soil, or alternatively the worm may fill up the interstices between crocks with soil to such an extent that no water can escape, when, of course, the soil becomes sour, and finally, bereft of proper food in the shape of decaying leaves, etc., the worm may in desperation attack the weakest and softest roots of the plant, which, however, are never touched except in cases of extremity.

When the sickly appearance of a pot plant indicates the presence of a worm, the best plan is to dissolve as much of a lump of lime as is soluble in water, and soak the soil with the lime water. The worm will quickly appear on the surface of the pot, and can be removed. This method must not be adopted with rhododendrons, azaleas, ericas and other plants which abhor lime, but in their case the plant may be removed from its pot, and allowed to stand an hour or two on a tiled or cemented floor. The worm will, in all probability, leave the root-ball during that period, or at any rate expose itself to view, when it can be extricated with the aid of a bent wire.

INDEX